DSP Implementation
using the TMS320C6000™ DSP Platform

Pearson
Education

DSP Implementation
using the TMS320C6000™ DSP Platform

Naim Dahnoun

An imprint of Pearson Education

Harlow, England · London · New York · Reading, Massachusetts · San Francisco
Toronto · Don Mills, Ontario · Sydney · Tokyo · Singapore · Hong Kong · Seoul
Taipei · Cape Town · Madrid · Mexico City · Amsterdam · Munich · Paris · Milan

Pearson Education Limited

Edinburgh Gate
Harlow
Essex CM20 2JE
England

and Associated Companies around the World

Visit us on the World Wide Web at:
www.pearsoneduc.com

First edition 2000

ISBN 0201-61916-4

British Library Cataloguing in Publication Data
A catalogue record for this book can be obtained from the British Library

Library of Congress Cataloging-in-Publication Data
Available from the publisher

10 9 8 7 6 5 4 3 2 1
04 03 02 01 00

Typeset by 43 in 10.5/13 Times
Printed and bound in Great Britain by T.J. International Ltd., Padstow, Cornwall

I dedicate this book to
Maria, Zahra, Yasmin and Riyad
and in memory of
Toufik and Rachid

Contents

Preface

Digital signal processing techniques are now so powerful that sometimes it is extremely difficult, if not impossible, for analogue signal processing to achieve the same or closer performance. Added to this, digital signal processors are very affordable and include good development tools and support. This is sufficient to explain the growing number of areas of application for DSP, including motor drives, communications, biomedical instrumentation and automotive applications.

Having dealt for some time with undergraduate and postgraduate students, researchers and digital signal processor users in general, I have found that first-time users of DSP find a barrier obstructing them in progressing from theory to the full implementation of algorithms.

When it comes to implementing an algorithm many questions arise, questions such as:

- Which processor to use – fixed or floating point?
- Which manufacturer to choose?
- Which application hardware to use?
- How many I/O interfaces are needed and how fast should they be?

When these questions are answered, more questions arise regarding the implementation on the specific processor and hardware selected. In this book, use of the TMS320C6000 will be justified, and the hardware and complete implementation of selected algorithms will be dealt with in detail. Material used for the teaching of undergraduate and postgraduate students, along with laboratory experiments, are used to demonstrate and simplify the transition from theory to the full implementation on the TMS320C6201 processor.

This book is divided into nine chapters. Chapters 2 and 3 are very important and it is advisable that they are well understood before progressing onto subsequent chapters.

Chapter 1 Introduction

This introductory chapter provides the reader with general knowledge on general-purpose DSP processors and also provides an up-to-date TMS320 roadmap showing the evolution of Texas Instruments' DSP chips in terms of processing power.

Chapter 2 The TMS320C62xx/C67xx architecture

The objective of this chapter is to provide a comprehensive description of the 'C6x architecture. This includes a detailed description of the Central Processing Unit (CPU) and program control along with an overview of the memory organisation, serial ports, boot function and internal timer.

Chapter 3 Software development tools and TMS320C6201 EVM overview

This chapter is divided into three main parts. The first part describes the software development tools, the second part describes the Evaluation Module (EVM) and, finally, the third part describes the codec, and use of interrupts along with some useful programs for testing the TMS320C6201 EVM.

Chapter 4 Software optimisation

To introduce the need for code optimisation, this chapter starts by developing the concept of pipelining. Since the TMS320C62xx and the TMS320C67xx each have eight units, which are dedicated to different operations, and since different instructions can have different latencies, the programmer or the tools are left with the burden of scheduling the code. Backed by examples, this chapter explains the different techniques used to optimise DSP code on these processors.

Chapter 5 Finite Impulse Response (FIR) filter implementation

The purpose of this chapter is twofold. Primarily, it shows how to design an FIR filter and implement it on the TMS320C62xx processor, and secondly, it shows how to optimise the code as discussed in Chapter 4. This chapter discusses the interface between C and assembly, how to use intrinsics, and how to put into practice material that has been covered in the previous chapters.

Chapter 6 Infinite Impulse Response (IIR) filter implementation

This chapter introduces the IIR filters and describes two popular design methods, that is the bilinear and the impulse invariant methods. Step by step, this chapter shows the procedures necessary to implement typical IIR filters specified by their transfer functions. Finally, this chapter provides complete implementation of an IIR filter in C language, assembly and linear assembly, and shows how to interface C with linear assembly.

Chapter 7 Adaptive filter implementation

This chapter starts by introducing the need for an adaptive filter in communications. It then shows how to calculate the filter coefficients using the Mean Square Error (MSE) criterion, exposes the Least Mean Square (LMS) algorithm and, finally, shows how the LMS algorithm is implemented in both C and assembly.

Chapter 8 Goertzel algorithm implementation

This chapter deals with Dual Tone Multi-Frequency (DTMF) detection and provides a practical example of the Goertzel algorithm. This chapter also shows how to produce optimised code by the pen and paper method, describes linear assembly and demonstrates how to program the Direct Memory Access (DMA).

Chapter 9 Implementation of the Discrete Cosine Transform

This chapter starts by introducing the need for video compression to reduce the channel bandwidth requirement, then explains the Joint Photographic Experts Group (JPEG) image codec. This includes a detailed discussion and the implementation of the Discrete Cosine Transform (DCT) and Inverse Discrete Cosine Transform (IDCT) and concentrates on their optimisation. An explanation of the PC–DSP communication via the PCI bus is also provided.

Software

The accompanying CD includes all the programs used in this book. To help the reader in locating or viewing the files, an Index.htm file has been included. The files are in separate directories corresponding to each chapter. Some directories are further divided in sub-directories to separate different implementations. Batch files for compiling, assembling and linking these programs are included. All the files have been tested (the environment may need to be modified: see env.bat file). Software using Code Composer Studio environment is also provided. Software updates including code running on the TMS320C6211 DSK can be obtained from the Publisher.

Acknowledgements

As you can imagine, it is hard to produce any textbook on state-of-the-art technology, especially when the time factor is playing against you. However, with the first very encouraging comments from the five anonymous reviewers, my motivation for writing this book surged, and therefore I would like to thank them for their constructive comments.

Due to the unfamiliarity with this processor, it was difficult to share ideas with other users. But with Tuan-Kiang Chiew, Kwee-Tong Heng and Michael Hart many problems were solved and many grey areas were clarified; I extend to them special thanks.

I am indebted to Robert Owen, Hans Peter Blaettel, Gene Frantz, Neville Bulsara, Greg Peake, Helga and Graham Stevenson and Maria Ho of Texas Instruments for their encouragement, continuous help and support. I owe my thanks to Professor Barrie Jones, Professor David Evans, Dr John Fothergill and Fernando Schlindwein from Leicester University for their encouragement, and Dr Anthony Brooms from Oxford University and Dr Mark Yoder from the Rose-Hulman Institute of Technology, USA, for reviewing the material.

My thanks to all of my colleagues at the Department of Engineering at Bristol University and to all of our students, in particular Khaled, Fernando, Mohamed, Samir, Chris, Shirley and Julian. Also I would like to thank Cornelius Kellerhoff, European DSP Business Development Consultant, Paul Coulton, Communications Research Centre, Lancaster University, and Mariusz Jankowski, University of Southern Maine, for their valuable technical reviews.

I thank my parents, family and friends for their support and encouragement.

Finally, many thanks to Karen Sutherland, Julie Knight and all of the Pearson Education and Prentice Hall team who were very kind, supportive and encouraging.

N. Dahnoun
Naim.Dahnoun@Bristol.ac.uk

Chapter 1

Introduction

1.1 Introduction

The Texas Instruments (TI) TMS320C6000 ('C62xx/'C67xx) are the latest and highest performance fixed and floating point DSP processors available from TI. These processors, which already operate at high clock rates of 300 MHz, will continue to see further increases in speed.

The combination of high speed and multiple units operating simultaneously has pushed the performance up to 2400 million instructions per second (2400 MIPS at 300 MHz). This means that these processors are 10 times faster than any other processor on the market at the present time. The 'C6xx are the first processors to benefit from a highly efficient C compiler and an assembly optimiser. The high performance of these processors makes them the first choice for high performance modem applications such as base stations, remote access servers (RAS), digital subscriber loops (DSL), cable modems and multichannel telephony systems. It is obvious that more applications will emerge in the future as clock speeds and internal memory increase and power consumption and cost decrease.

Before embarking on the study of the 'C62xx/'C67xx processors, a few questions are worth answering and these are dealt with in Sections 1.2–1.5 below.

1.2 Why do we need DSP processors?

Traditional signal processing was achieved by using analogue components such as resistors, capacitors and inductors. However, the inherent tolerance associated with these components, temperature and voltage changes, and mechanical vibrations, can dramatically affect the effectiveness of analogue circuitry. On the other hand, digital signal processing is inherently stable, reliable and repeatable. In addition to this the advancement in VLSI has opened up a new horizon in real-time digital signal processing and enabled applications that were not possible using analogue techniques, for example linear phase filters. With DSP it is easy to change, correct or update applications and additionally DSP reduces noise susceptibility, chip count, development time, cost and power consumption.

DSP processors are similar to general-purpose microprocessors except that they are more optimised to perform multiplication and addition operations. DSPs also have the advantage of consuming less power and being relatively cheap. The first low cost DSP device, the TMS320C10, was introduced by Texas Instruments in 1982.

1.3 How do we define real-time?

The definition of real-time depends on the application; for instance, if we consider an audio application that requires a sampling frequency of 40 kHz, needing 100 serial instructions to complete and using a typical DSP processor with 30 ns cycle time, then in 3 μs (30 ns $\times$ 100) the calculation is completed, whereas the cycle time is 25 μs. Finally, we can say that we have a real-time application as long as the waiting time is greater than or equal to zero (see Figure 1.1).

1.4 What are the typical DSP algorithms?

Digital signal processing algorithms are inexhaustible. However, the Finite Impulse Response (FIR) filters, Infinite Impulse Response (IIR) filters,

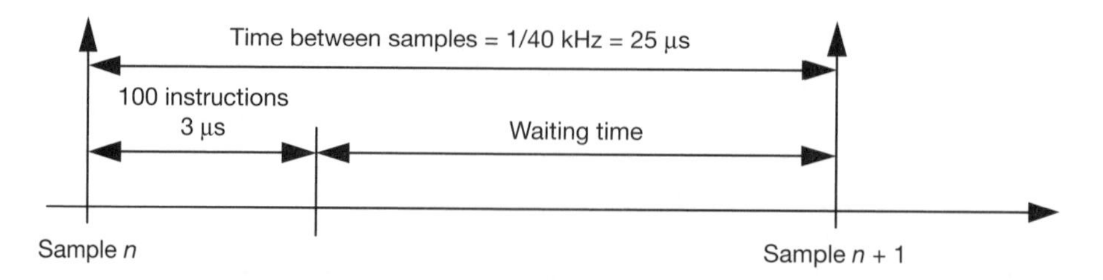

Figure 1.1 Definition of real-time

Algorithm	Formula
FIR	$y(n) = \sum_{k=0}^{M} a_k x(n-k)$
IIR	$y(n) = \sum_{k=0}^{M} a_k x(n-k) + \sum_{k=1}^{N} b_k y(n-k)$
Conv.	$y(n) = \sum_{k=0}^{N} x(k)h(n-k)$
DFT	$X(k) = \sum_{n=0}^{N-1} x(n) \exp[-j(2\pi/N)nk]$
DCT (8 × 8)	$\frac{1}{4}\sum_{x=0}^{7}\sum_{y=0}^{7} c(u).c(v)f(x,y).\cos[\pi(2x+1)u/16].\cos[\pi(2y+1)v/16]$

Table 1.1 Common DSP algorithms

Convolution, Discrete Fourier Transforms (DFT) and Discrete Cosine Transforms (DCT) are amongst the most common DSP algorithms. From Table 1.1 it is clear that all these algorithms share common operations such as multiply and accumulate. These operations form the key features of DSP processors, since they are performed in hardware and can be effectively executed in a single processor cycle.

1.5 Who are the main general-purpose DSP processor manufacturers?

Table 1.2 shows the main manufacturers of general-purpose DSP processors currently available and their associated web sites. Updates can be found on the appropriate web sites shown in the table. Due to the very fast evolution of DSP processors, companies such as BDTI (www.bdti.com) and Forward Concept (www.fwdconcepts.com) have been set up to help in choosing the right processor by supplying independent reports and benchmarks for various DSP algorithms and processors. Figure 1.2 shows the DSP market share among the main suppliers. However, this may change in the future. At the time of writing, Motorola and Lucent, and Analog Devices and Intel were both in the process of pooling resources for joint DSP investments.

DSP supplier	Device name	Fixed point	Floating point	Word length (bits)*
Texas Instruments	TMS320C1X	•		16
at www.ti.com	TMS320C2X	•		16
	TMS320C2XX			16
	TMS320C3X		•	32
	TMS320C4X		•	32
	TMS320C5X	•		16
	TMS320C54X	•		
	TMS320C62XX	•		16
	TMS320C67XX		•	16
	TMS320C8X	•	•	32
Lucent	ADSP 16x/16xxx	•		16
at www.lucent.com	ADSP32C/3210		•	32
Motorola	DSP56000	•		16
at www.mot.com	DSP56100	•		16
	DSP56300	•		24
	DSP56600	•		24
	DSP56800	•		16
	DSP96002		•	32
Analog Devices	ADSP-218x	•		16
at www.analog.com	ADSP-219x	•		16
	ADSP-2106x		•	32
	ADSP-2116X		•	32
	Tiger SHARC	•	•	8/16/32

* Based on the multipliers' operands

Table 1.2 Main general-purpose DSP manufacturers and their products

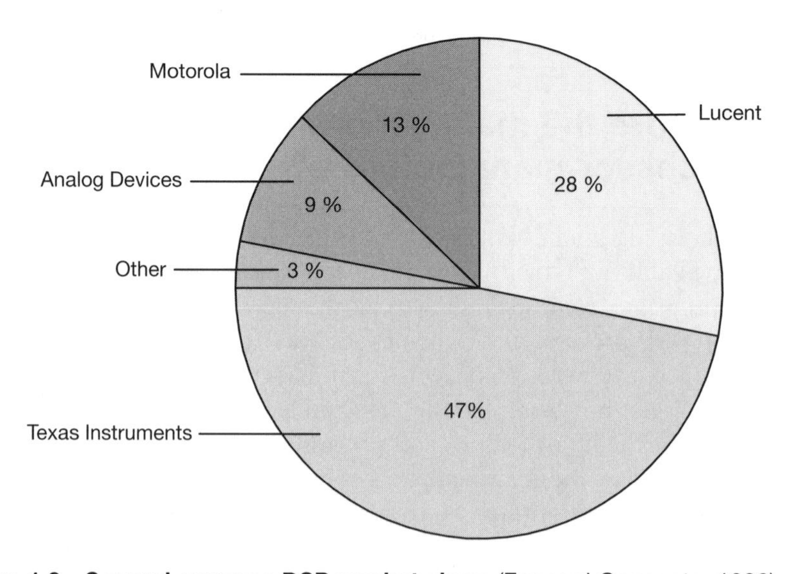

Figure 1.2 General-purpose DSP market share (Forward Concepts, 1998)

1.6 Parameters to be considered when choosing a DSP processor

Various parameters have to be considered when selecting a suitable device (see Table 1.3). However, selecting a suitable device on its own is not always sufficient, and selecting a suitable DSP platform is as important as selecting a DSP device. In Table 1.3 the parameters are divided into two sections, one specific for a device and the other related to systems. Table 1.3 also shows parameters specific to the 'C6201 and 'C6701 processors taken as examples. Applications which require high precision, wide dynamic range,

Device parameters	TMS320C6201 (at 200 MHz)	TMS320C6701 (at 167 MHz)
Arithmetic format	32-bit	32-bit
Extended floating point format	NA	64-bit
Extended arithmetic format	40-bit arithmetic	40-bit arithmetic
Performance (peak and typical)	(1600–?*) MIPS	(1336–?*) MFLOPS
External memory bandwidth	880 Mbits/s	
I/O bandwidth:		
Serial ports (number and speed)	2 × 50 Mbits/s	2 × 83.4 Mbit/s
Host port:	[spra449a.pdf]	[spra449a.pdf]
Read with auto-increment	up to 277.6 Mbits/s	up to 231.79Mbits/s
Write with auto-increment	up to 400 Mbits/s	334 Mbits/s
Number of hardware multipliers	2 × (16 × 16-bit) with 32-bit result	2 × (32 × 32-bit) with 32- or 64-bit result
Number of registers	32	32
Internal program memory size	16K × 32-bit	16K × 32-bit
Internal data memory size	16K × 32-bit	16K × 32-bit
Cache	Entire program memory if selected	Entire program memory if selected
DMA channels	4 + 1 auxiliary	4 + 1 auxiliary
Multiprocessor support	Not inherent	Not inherent
Power consumption	< 2 watts	< 2 watts
Power management	Yes	Yes
On-chip timers (number and width)	2 × 32-bits	2 × 32-bits
Cost (in $) (sample pricing)	$160	$257
Size (W × H × D) and package	Refer to data sheet	Refer to data sheet
External Memory Interface Controller	Available	Available
JTAG	Available	Available

* Typical values depend on the application

Table 1.3 (a) DSP device parameters to be considered for choosing a processor or platform

System parameters	TMS320C6201 (at 200 MHz)	TMS320C6701 (at 167 MHz)
Number of I/O modules	These parameters depend on third parties supplying DSP platforms	
PCI, ISA, VME, etc., interface		
Cost		
Development tools		
Technical support		
Failure test		
Upgradable		
Power consumption		
Application reports		

Table 1.3 (*cont.*) **(b) DSP system parameters to be considered for choosing a processor or platform**

high signal-to-noise ratio and ease of use obviously require a floating point device. It is the application that dictates which device and platform to use in order to achieve optimum performance at the lowest cost.

1.7 General-purpose DSP vs DSP in ASIC

An Application Specific Integrated Circuit (ASIC) is a semiconductor designed for a dedicated function. DSP ASIC libraries, which include arithmetic units, control units, memory units and buses, are available from companies such as Analog Devices, Texas Instruments and Integrated Silicon Systems Ltd. Although DSP processors in ASICs offer single-chip solutions, are small in size, run faster and consume less power (see Table 1.4), they do require a large investment cost and are only considered if mass quantity (several thousand units) is required.

Advantages	Disadvantages
• High throughput	• High investment cost
• Lower silicon area	• Less flexibility
• Lower power consumption	• Long time from design to market
• Improved reliability	
• Reduction in system noise	
• Low overall system cost	

Table 1.4 Advantages and disadvantages of an ASIC

1.8 DSP market

Although a big share of the DSP market is in control and specifically in motor drive applications, DSP applications in communications are amongst the fastest growing areas. In fact the global DSP market is growing 50% faster than the remaining semiconductor market. The general-purpose DSP market totalled US$3.5 billion in 1998 and will reach $12 billion by 2002, according to Forward Concepts.

1.9 The TMS320 family evolution

The TMS320 family has reached its ninth generation in order to satisfy the need for continuous high performance DSP processors (see Figure 1.3). The TMS320 family can be divided into three groups of processors, namely the fixed point processors ('C1x, 'C2x, 'C2xx, 'C5x, 'C54x and 'C62xx), the floating point processors ('C3x, 'C4x and 'C67xx), and multiprocessors ('C8x). The 'C8x family can have up to five processors (one processor is floating point and the rest are fixed point) running in parallel. In each generation, various processors are available, each sharing the same CPU core but having different peripherals, thus reducing cost and power consumption without compromising performance.

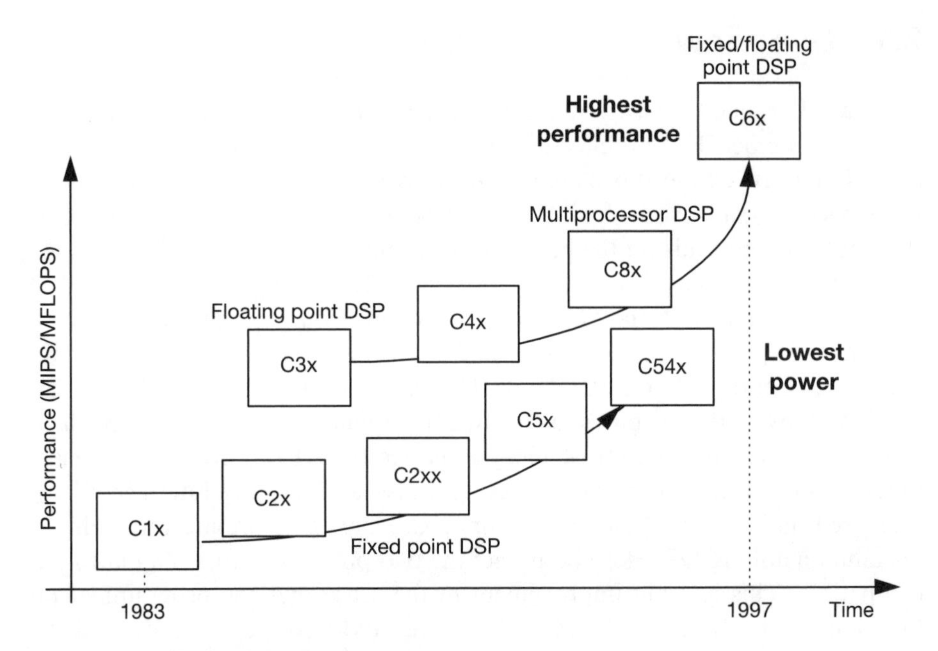

Figure 1.3 TMS320 family evolution

Chapter 2

The TMS320C6000 architecture

2.1 Overview

The TMS320C6000 devices are the first DSP chips of a new generation based on the TI VelociTI™ architecture, which is an enhancement of the VLIW (Very Long Instruction Word architecture). By keeping the architecture simple, the device is made very flexible. Whether the processor is easy to use is debatable and depends on the performance required.

The 'C62xx are fixed point processors currently running at a clock speed of up to 300 MHz and the 'C67xx are floating point processors. As with conventional processors, the 'C62xx and 'C67xx are composed of three main parts: the Central Processing Unit (CPU), memories and peripherals, all connected by internal buses as shown in Figure 2.1. In addition these processors have an External Memory Interface (EMIF) to provide a glueless interface to common memory devices, and also a Host Port Interface (HPI). The fixed and floating point processors share the same architecture. This is uncommon among DSP devices; in fact the two processors are even pin-to-pin compatible. This is a big improvement in the sense that the programmer no longer needs to understand two architectures and two sets of tools, and the DSP platform manufacturers need only to develop a single platform for both

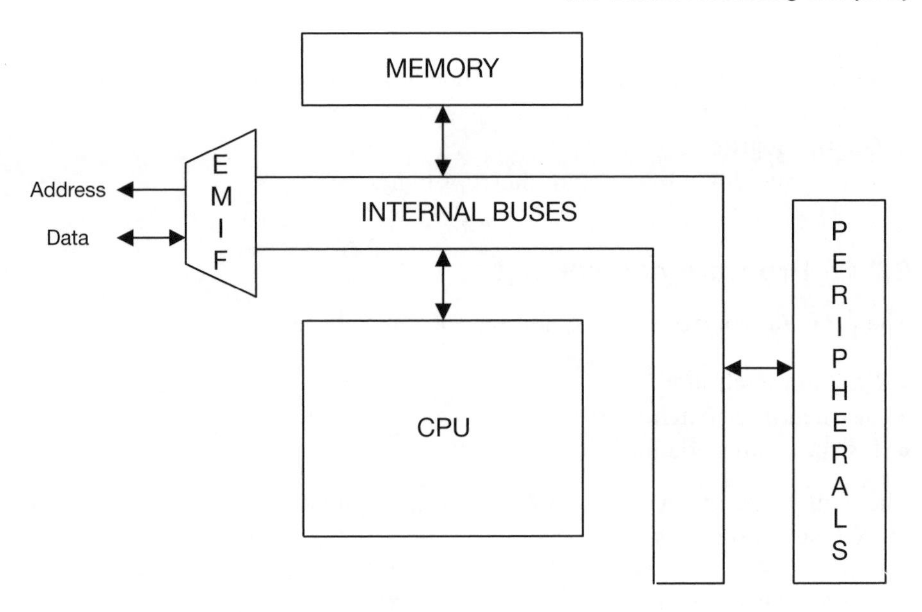

Figure 2.1 TMS320C6000 block diagram

types of processors. Additionally, identical architectures also have the advantage of code conversion from fixed to floating point and vice versa with minimum programming effort.

2.2 The Central Processing Unit (CPU)

The CPU, which is the heart of the processor, is composed of the following elements (Figure 2.2). Each element is described separately below.

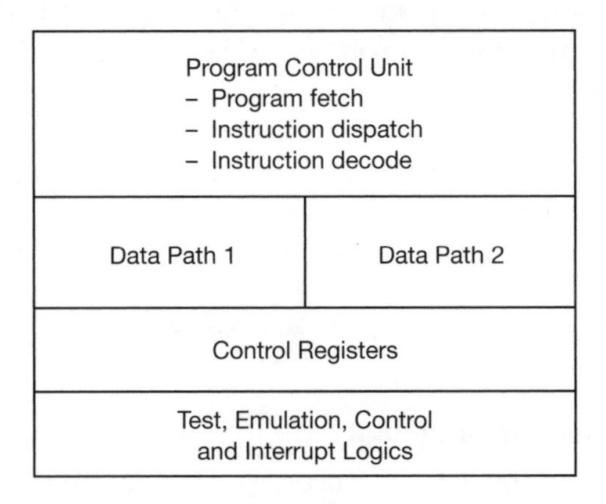

Figure 2.2 TMS320C6000

- Program control unit
- Two data paths (each with four functional units and 16 32-bit general-purpose registers)
- Control registers
- Test, emulation, control and interrupt logics.

2.2.1 Program control unit

The program control unit is composed of three elements:

- Program fetch unit
- Instruction dispatch unit
- Instruction decode unit.

These units operate in an assembly line fashion and are necessary for the CPU to execute instructions.

> Note: Each unit is capable of handling up to eight instructions per cycle.

2.2.1.1 Program fetch unit

To retrieve a fetch packet (FP), which is a group of eight instructions, four phases are required.

- PG phase: the CPU generates a fetch address
- PS phase: the CPU sends the address to the memory
- PW phase: the CPU waits for the data to be ready
- PR phase: the CPU reads the opcode.

The program address of the first instruction in the fetch packet is generated in the PG phase (cycle n) and sent to the program memory in the PS phase (cycle $n+1$). The CPU waits for the instructions (all eight instructions, whether in parallel or not) in the program memory to be become available (PW phase). This will take one cycle if the program is in the internal program memory or eight cycles if the program is in the external memory with zero wait state, otherwise it will take $(8+m)$ cycles if the external program memory has m wait state(s). In the last phase of the fetch unit, which is the PR phase, the CPU receives the fetch packet. This is illustrated in Figure 2.3.

> Note: A fetch packet (FP) is a group of eight instructions.

2.2.1.2 Instruction dispatch unit

Since the 'C62xx and the 'C67xx processors each have eight functional units and each functional unit can only execute specific instructions, the instructions

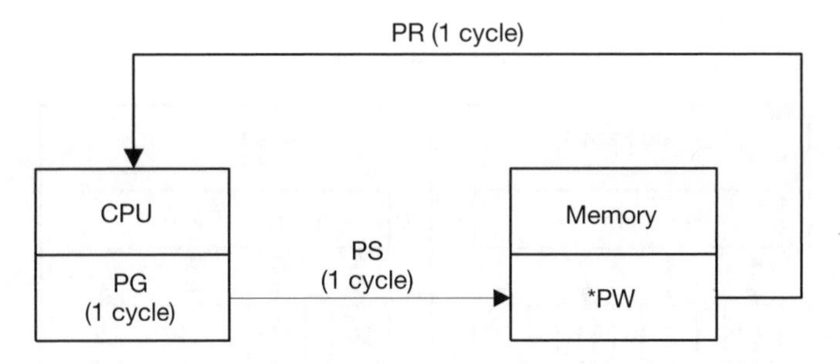

Figure 2.3 Illustration of the program fetch phases

have to be dispatched to the appropriate units. This is achieved by the instruction dispatch unit, Figure 2.4.

2.2.1.3 Instruction decode unit

When the instruction opcode reaches the instruction decode unit, the opcode is decoded. Although the instruction decode unit DC is shown as a separate unit in Figure 2.4, the actual operation is performed by the functional units.

2.2.2 CPU data paths

The 'C62xx and the 'C67xx CPUs are composed of two blocks known as data path 1 and data path 2, as shown in Figure 2.5. Each block has four execution units known as .L, .M, .S and .D, a register file containing 16 32-bit general-purpose registers, and multiple paths for: (1) data communications between each block and memory, or (2) data communications within each block, or (3) data communication between blocks.

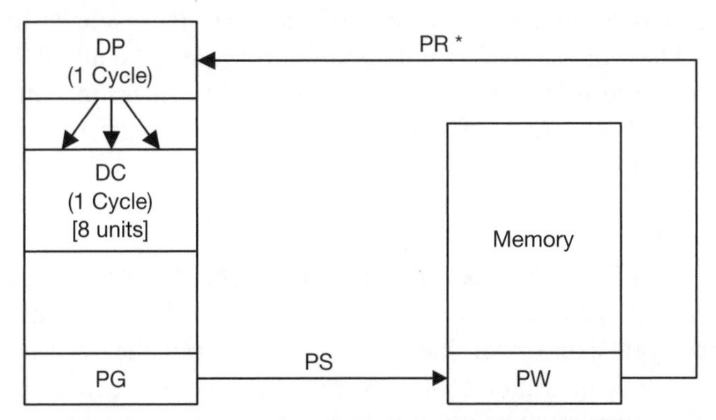

Figure 2.4 Illustration of data flow

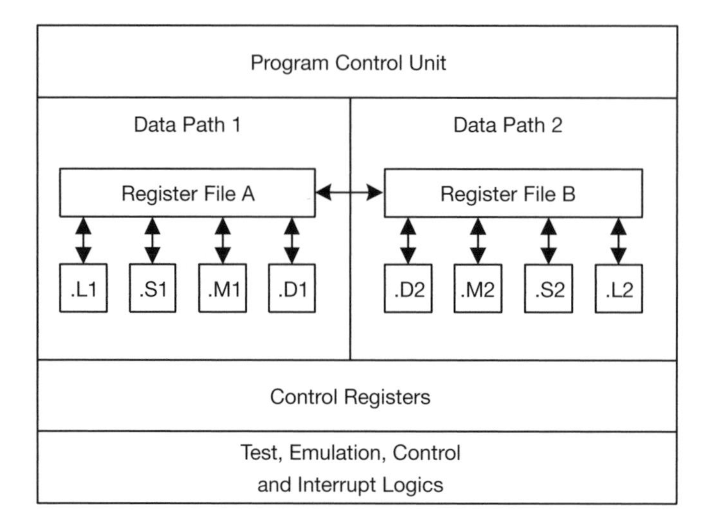

Figure 2.5 'C62xx and 'C67xx CPU block diagram

From Figure 2.6 (referring to the 'C62xx) and Figure 2.7 (referring to the 'C67xx) it can be seen that register file A can be written to or read from functional units .L1, .S1, .M1 and .D1 via the paths indicated by arrows. The same can be applied to register file B where all registers can be accessed by functional units .L2, .S2, .M2 and .D2. The CPU paths can be divided into two types, one being the data path and the other the address path. The data paths are used for data transfer between the register files and the units, or for data transfer between the memory and the register files. However, the address path is used for sending the address from the data unit .D to the memory.

2.2.2.1 Cross paths

Cross paths enable linking of one side of the CPU (e.g. side 1) to the other (e.g. side 2). These are shown by bold arrows in Figures 2.6 and 2.7. Although the cross paths are useful in terms of the flexibility in using units with two operands from both sides of the CPU, there are restrictions which are discussed in the following sections.

Data cross paths

The data cross paths can also be referred to as the register file cross paths. These cross paths allow operands from one side to cross to the other side. There are only two cross paths: one from side B to side A (1X) and one from side A to side B (2X). These limit the number of cross paths to two for each execute packet. The following points must be observed:

- Only one cross path per direction per execute packet is permitted.
- The destination register is always on the same side of the unit used.

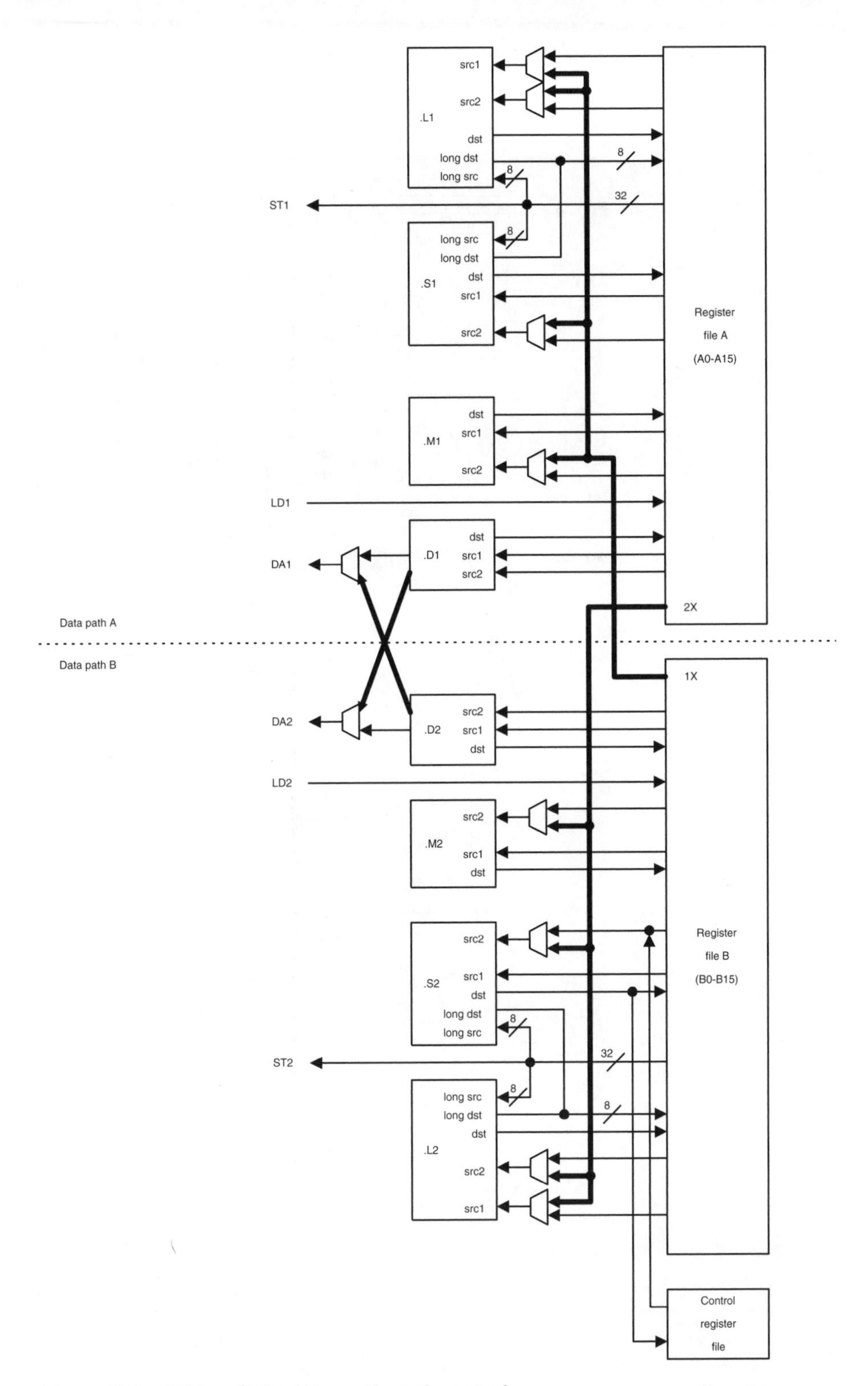

Figure 2.6 'C62xx CPU data path and control

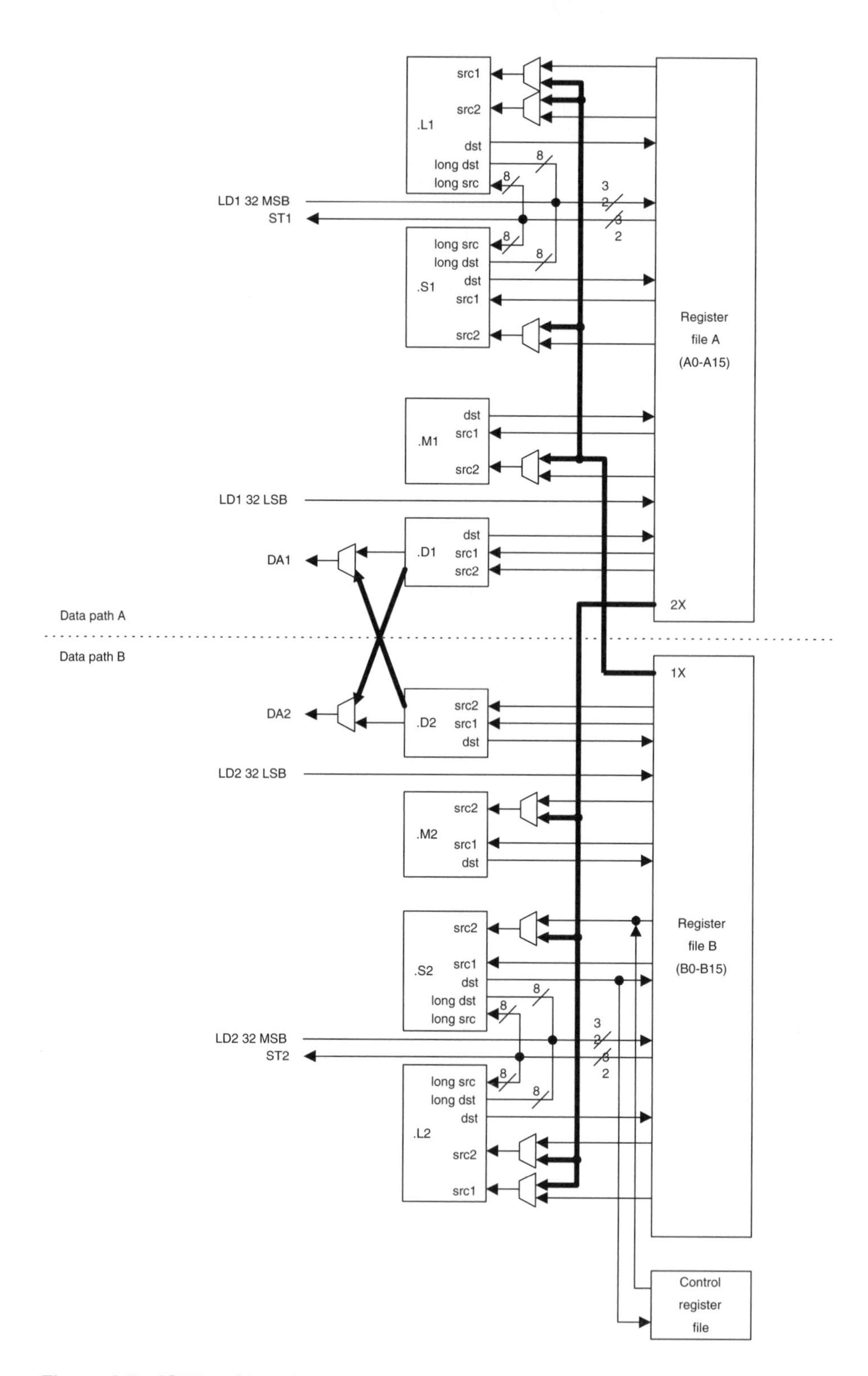

Figure 2.7 'C67xx CPU data path and control

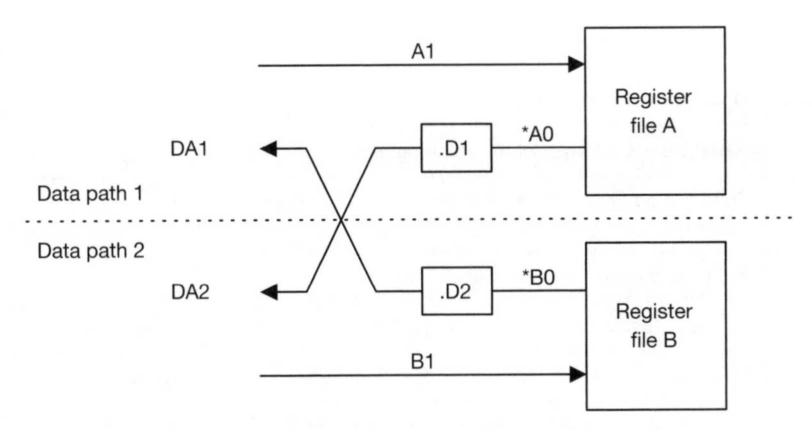

Figure 2.8 Address cross paths

Address cross paths

The addresses generated by the data units .D1 and .D2 can be sent to either the data address path DA1 or the data address path DA2, as shown by bold arrows in Figures 2.6 and 2.7. The advantage of using an address cross path is to be able to generate the address using one register file, and accessing the data from the other register file as shown in Figure 2.8. Here again there are only two cross paths for each execute packet and the following points should be observed:

- Only one address cross path per direction per execute packet is allowed.
- When an address cross path is used, the destination register for the load (LD) instructions and the source register for the store (ST) instruction should come from the opposite side of the unit (see Figure 2.8), or simply the register pointers must come from the same side of the .D unit used.
- If both .D units are to be used, then either none or both of the address cross paths should be used.

2.2.3 Program execute units

In these units (four units in each data path) the instructions are finally executed. Although in the execute unit there are five phases, not all instructions go through all of these phases. Some instructions finish executing in the initial phase (E1) (see Table 2.1) and are called single-cycle instructions (SCI), whereas other instructions finish executing in phase two (E2) which are the multiply instructions. The store instruction finishes in phase three (E3) whereas the load finishes in phase five (E5). For the branch instruction the branch target address is generated in E1, and the target code has to go through the pipeline stages (PG, PS, PW, PR, DP and DC) before reaching E1. This will result in the branch taking six cycles. For the 'C67xx processors the principle

Phase	Operation(s) in this phase		Type of instructions to be completed in this phase
PG	Generate the address of the fetch packet		None
PS	Send the address of the fetch packet to the program memory		None
PW	Program memory is accessed		None
PR	The fetch packet is received by the CPU		None
DP	Dispatch the instruction(s) to the functional unit(s)		None
DC	The instructions are decoded in the appropriate functional units		None
E1	All instructions	The conditions are evaluated Operands are read	All instructions *except* Multiply, Store and Load
	Load & Store	If the conditions are true, then (1) Address generation is performed (2) Address modifications are updated to registers*	
	SCI	Executions are completed and result written to the registers	
	Branch	The branch address in the PG stage is modified according to the branch operand, in order to load the branch target code	
E2	Load	The address is sent to the memory	Multiply
	Store	The address and data are sent to the memory	
	SCI	SAT bit in the Control Status Register is set if saturation has occurred	
	Multiplication	Results of the multiplications are written to the registers	
E3	Store	The data is stored in the memory	Store
	Multiplication	SAT bit in the Control Status Register is set if saturation has occurred due to a multiplication	
E4	Load	The data is received by the CPU	None
E5	Load	The data is written to a register	Load

* If the conditions are *not* true, no further operations on any instruction are performed

Table 2.1 Pipeline phases for the 'C62xx processors

is the same; however, the execute unit has 10 phases (TMS320C62xx/C67xx *CPU and Instruction Set Reference Guide*, SPRU189).

There are four execution units for each data path. All the units operate on 32-bit operands and execute instructions simultaneously. However, the .L and .S units can also operate on 40-bit operands. Each unit executes a specific set of operations, which can be seen as a limitation; for example, there are only two multiplications possible per cycle.

2.2.3.1 .L Units

These are 40-bit integer Arithmetic and Logic Units (ALUs). These two units (.L1 and .L2) can be used for:

- 32/40-bit arithmetic and compare operations
- 32-bit logical operations
- Normalisation and bit count operations
- Saturated arithmetic for 32/40-bit operations.

2.2.3.2 .M Units

There are two hardware multiplier units, .M1 (for data path 1) and .M2 (for data path 2). These units are capable of performing 16-bit by 16-bit multiplications producing 32-bit results (for the 'C62xx). Because the input operands to the multipliers come from the 32-bit registers, there are several variants of the multiply instruction in order to avoid data manipulation on the registers or extra loads. As shown in Table 2.2, the operands may come from the 16 MSB or the 16 LSB of the registers used and therefore four variants of the multiplication instruction are available.

As stated earlier, the instructions load, store, multiply and branch have different latencies and therefore complicate programming. In the case of multiplication (only for the 'C62xx) the latency is two instruction cycles. However, when instructions are pipelined the multiplier can issue one instruction per cycle. The multiplier units support all combinations of signed and unsigned operands for all multiplication variants and also support multiplication with left shift and saturation (SMPY). (Refer to the *CPU and Instruction Set Reference Guide* for more details on each multiplication instruction.)

		Register a			Register b			Register c
MPY	a,b,c		a	×		b	=	a × b
MPYH	a,b,c	A		×	B		=	A × B
MPYHL	a,b,c	A		×		b	=	A × b
MPYLH	a,b,c		a	×	B		=	a × B

Table 2.2 Different multiplication instructions

2.2.3.3 .S Units

These units (.S1 and .S2) contain 32-bit integer ALUs and 40-bit shifters. These units can be used for:

- 32-bit arithmetic, logic and bit field operations
- 32/40-bit shifts
- Branches (.S2 only when using a register)
- Register transfers to and from control registers (***.S2 only***)
- Constant generation.

Note: All instructions executing in the .L or .S are single-cycle instructions, except for the branch instructions.

2.2.3.4 .D Units

The data units (.D1 and .D2) can be used for the following operations:

- Load and store with 5-bit constant offset
- Load and store with 15-bit constant offset (***.D2 only***)
- 32-bit additions/subtractions
- Linear and circular address calculations.

2.2.4 Control registers

The 'C62xx devices have 10 registers for control purposes, while the 'C67xx have 13 control registers, as shown in Table 2.3. The three extra registers available on the 'C67xx processors are to support floating point operations. Eight registers are dedicated for interrupt control and are described in detail in Section 2.8. As shown in Figures 2.6 and 2.7, reading and writing to the control registers can only be performed via the .S2 unit. All the control registers can only be accessed by the MVC (move constant) instruction.

Note: Only the .S2 unit and the MVC instruction can be used to access the control registers.

2.2.5 Register files

Each data path contains a register file composed of 16 32-bit general purpose registers (A0–A15 for data path 1 and B0–B15 for data path 2). These registers can support 32- and 40-bit fixed point data or 64-bit double-precision floating point data for the case of the 'C67xx. Only the .L and .S units support these extended operations. These general-purpose registers can be used for:

- data
- data address pointers, or

Control register	Brief description
Addressing mode register (**AMR**)	• Used for specifying circular or linear addressing modes • Holds the block sizes for the circular addressing
Control status register (**CSR**)	Contains status and control bits
Interrupt flag register (**IFR**)	Contains status bits for each interrupt
Interrupt set register (**ISR**)	Used for setting pending interrupts manually
Interrupt clear register (**ICR**)	Used for clearing pending interrupts manually
Interrupt enable register (**IER**)	Used for setting active interrupts
Interrupt service table pointer (**ISTP**)	Contains the address of the interrupt vector
Interrupt return pointer (**IRP**)	Contains the address to return to after a mask interrupt has been serviced
Nonmaskable interrupt return pointer (**NRP**)	Contains the address to return to after a non-maskable interrupt has been serviced
Program counter, E1 phase (**PCE1**)	Contains the address of the fetch packet in the E1 stage
Floating point adder configuration register (**FADCR**) ('C67xx only)	Contains two specific fields, one for .L1 and the other for .L2. These fields specify underflows, overflows, rounding modes, scr1 or scr2 as denormalised number inexact results and infinity results
Floating point auxiliary configuration register (**FAUCR**) ('C67xx only)	Similar to FADCR except that it applies to .S units instead of .L units
Floating point multiplier configuration register (**FMCR**) ('C67xx only)	Similar to FADCR except that it applies to .M units

Table 2.3 TMS320C6000 control registers

- conditional registers (only A2, A1, B0, B1 and B2 registers can be used as conditional registers).

To create 40- or 64-bit operands, two registers have to be concatenated and the registers must be:

- from the same data path, and
- ordered as odd (MSB) first and even (LSB) second (see Table 2.4).

The following sequence of code gives an example of how to use 40-bit operands:

```
ADD   .L1   A0 , A3:A2 , A5:A4    ; Add 32-bit operand with 40-bit operand and
                                  ; put the results in 40-bit registers. The add
                                  ; operation is using the .L1 functional unit.
```

Register file A	Register file B
A1:A0	B1:B0
A3:A2	B3:B2
A5:A4	B5:B4
A7:A6	B7:B6
A9:A8	B9:B8
A11:A10	B11:B10
A13:A12	B13:B12
A15:A14	B15:B14

Table 2.4 Possible 40/64-bit register pair combinations

It is possible to use cross paths with instructions using 40-bit operands as long as the operands from the cross paths are 32-bit, as the cross paths are only 32 bits wide. As an example consider the following code:

```
ADD   .L1x   A1:A2 , B2 , A3:A4    ; add 40-bit operand to 32-bit operand which is
                                   ; located on the opposite side of the .L1 unit.
                                   ; The 1x means that the cross-path from side 2
                                   ; to side 1 is used.
```

2.3 Memory

The 'C62xx/'C67xx devices have a total memory address range of 4 Gbytes (2^{32} since the address bus is 32 bits wide) divided into four spaces which are the internal program memory, internal data memory, internal peripheral and external memory map spaces (CE0, CE1, CE2, CE3). The exact location of these spaces depends on the type of memory map used (MAP 0 or MAP 1) as shown in Figure 2.9a. For some devices the internal program memory can be used for either program or cache, but not both. Since most applications used in this book are based on the 'C6201 evaluation module (EVM), the memory maps for the EVM are also shown (see Figure 2.9b) with the EVM BOOTMODE[4:0] pins set to select MAP 1.

The external memory spaces CE0, CE2 and CE3 support asynchronous (SRAM and ROM) and synchronous (SBSRAM and SDRAM) memory with 8- and 16-bit read/write, and 32-bit read only. The space CE1 supports 32-bit read/write, or 8- and 16-bit read only memory (see Table 2.5).

2.3.1 Data memory access

The CPU and DMA Controller requests are made through the Data MEMory Controller (DMEMC) as shown in Figure 2.10. The DMA controller only

External memory space	Memory type supported	Accessibility
CE0 CE2 CE3	• Asynchronous (SRAM and ROM) • Synchronous (SBSRAM, SDRAM)	• 8/16-bit Read/Write • 32-bit Read only
CE1	• Asynchronous	• 8/16-bit Read only • 32-bit Read/Write

Table 2.5 External memory supported

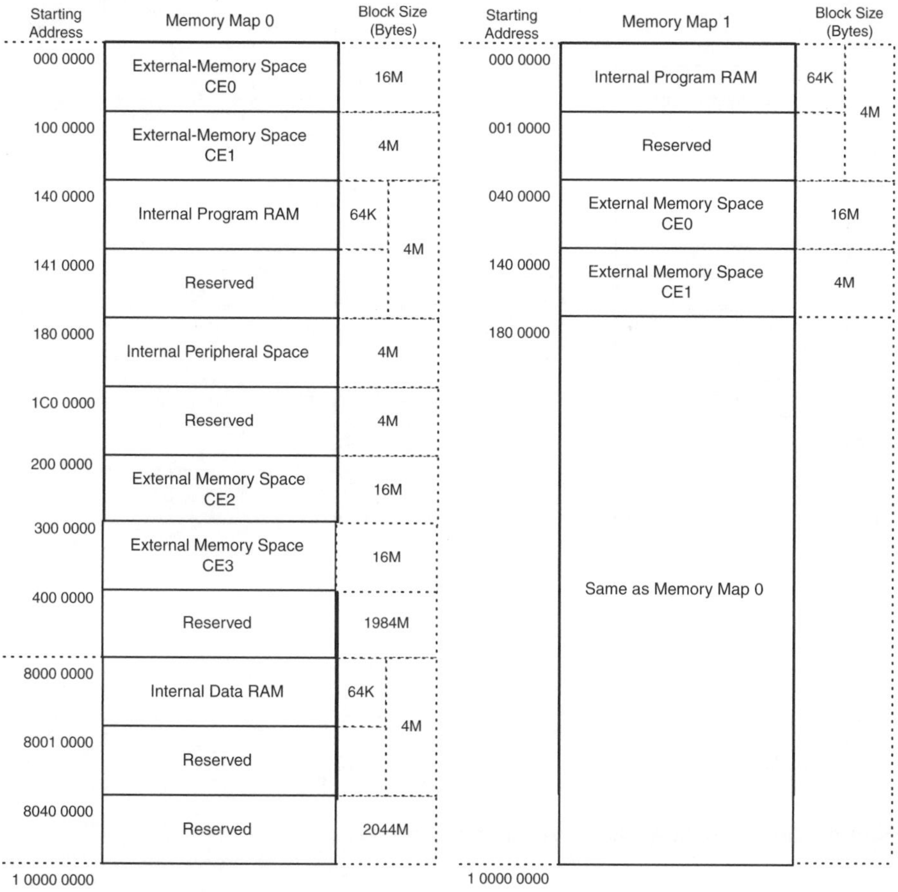

Starting Address	Memory Map 0	Block Size (Bytes)
000 0000	External-Memory Space CE0	16M
100 0000	External-Memory Space CE1	4M
140 0000	Internal Program RAM	64K (4M)
141 0000	Reserved	
180 0000	Internal Peripheral Space	4M
1C0 0000	Reserved	4M
200 0000	External Memory Space CE2	16M
300 0000	External Memory Space CE3	16M
400 0000	Reserved	1984M
8000 0000	Internal Data RAM	64K (4M)
8001 0000	Reserved	
8040 0000	Reserved	2044M
1 0000 0000		

Starting Address	Memory Map 1	Block Size (Bytes)
000 0000	Internal Program RAM	64K (4M)
001 0000	Reserved	
040 0000	External Memory Space CE0	16M
140 0000	External Memory Space CE1	4M
180 0000	Same as Memory Map 0	
1 0000 0000		

Figure 2.9a 'C6201 memory maps

Starting Address	EVM Memory Map 0	Block Size (Bytes)	
000 0000	SBSRAM	256K	16M
004 0000	Unused		
100 0000	Asynchronous Expansion Memory	3M	4M
130 0000	PCI add-on registers	64	64K
130 0040	Unavailable		
131 0000	PCI FIFO	4	64K
131 0004	Unavailable		
132 0000	Audio Codec Registers	16	64K
132 0010	Unavailable		
133 0000	Reserved	320K	
138 0000	DSP control/status registers	32	64K
138 0020	Unavailable		
139 0000	Reserved	448K	
140 0000	Internal Program RAM	64K	4M
141 0000	Reserved	Reserved	
180 0000	Internal Peripheral Space	4M	
1C0 0000	Reserved	4M	
200 0000	SDRAM (Bank 0)	4M	
240 0000	Reserved	12M	
300 0000	SDRAM (Bank 1)	4M	
340 0000	Reserved	12M	
400 0000	Reserved	1984M	
8000 0000	Internal Data RAM	64K	4M
8001 0000	Reserved		
8040 0000	Reserved	2044M	
1 0000 0000			

Starting Address	EVM Memory Map 1	Block Size (Bytes)	
000 0000	Internal Program Memory	64K	4M
001 0000	Reserved		
040 0000	SBSRAM	256K	16M
044 0000	Unused	Unused	
140 0000	Asynchronous Expansion Memory	3M	4M
170 0000	PCI add-on registers	64	64K
170 0040	Unavailable		
171 0000	PCI FIFO	4	64K
171 0004	Unavailable		
172 0000	Audio Codec Registers	16	64K
172 0010	Unavailable		
173 0000	Reserved	320K	
178 0000	DSP control/status registers	32	64K
178 0020	Unavailable		
179 0000	Reserved	448K	
180 0000	Same as EVM Memory Map 0		
1 0000 0000			

Figure 2.9b TMS320C6201 EVM memory maps

Data read and write requests using the DMEMC	
From CPU to:	**From DMA to:**
Internal data memory	Internal data memory
On-chip peripherals	
EMIF	

Table 2.6 CPU/DMA request sources when using the DMEMC

uses the DMEMC for internal data memory access. However, the CPU uses the DMEMC for data request to internal data memory as well as to on-chip peripherals (via the peripheral bus controller) or to the External Memory InterFace (EMIF) (see Table 2.6 and Figure 2.10).

The operation of data memory access is as follows. The CPU sends the access requests to the data memory controller by sending control signals and data addresses (through DA1 and DA2 buses) and then reads (through LD1 and LD2 buses) or writes (through ST1 and ST2 buses) via the DMEMC as shown in Figure 2.10. For the CPU/DMA access, the arbitration is performed by the DMEMC. See Chapter 8 for more details on the DMA.

2.3.2 Internal memory organisation for the 'C6201 Rev. 2

Memory organisation and peripherals are very much dependent on silicon revision. The applications in this book are based on the evaluation platform

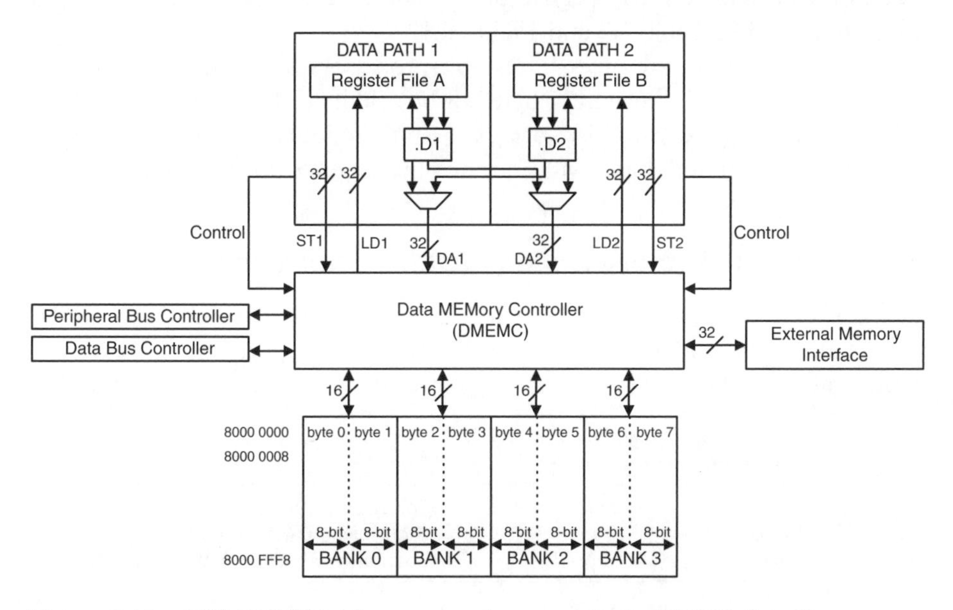

Figure 2.10 CPU-DMEMC Memory bank connections ('C6201 Rev. 2)

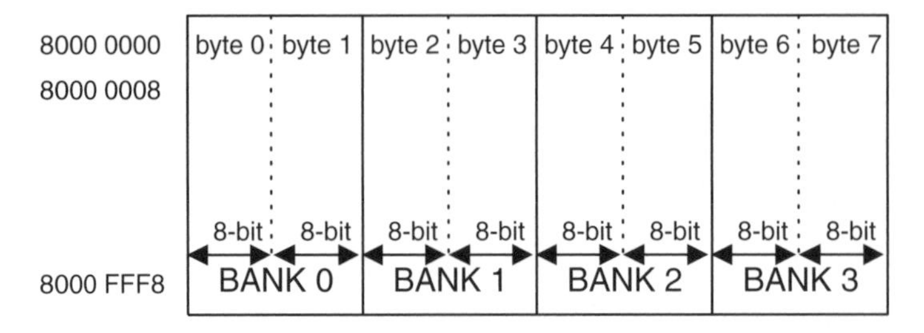

Figure 2.11 Memory bank arrangement

using the 'C6201 device revision 2 silicon and therefore only this specific device is dealt with in this section.

The memory organisation is also different for fixed and floating point processors and for applications on a specific processor and revision, the reader is advised to refer to the specific data sheets.

The internal data memory for the 'C6201 is organised in one block of four banks (BANK0–BANK3). Each bank is composed of 8K × 16-bit (Figure 2.11). The block of memory has been divided this way to allow the two data paths to load or store 16-bit data simultaneously. Therefore data accesses to different banks do not result in conflicts. However, if two simultaneous accesses to the same bank occur, the result is a memory conflict and the CPU will stall for an extra cycle. In conflicts involving two loads, the load using the DA2 bus will occur first, while for two stores, the store using DA1 will occur first; finally, if both load and store occur simultaneously, the load always occurs first irrespective of the data address path. These priorities are summarised in Table 2.7 and Figure 2.12.

The 'C62xx/'C67xx are byte-addressable processors, which means that both the CPU and DMA can access bytes (8-bit), halfwords (16-bit) and words (32-bit) (the 'C67xx can also access 64-bit – double words). However, accessing a byte in any bank will prevent simultaneous access to this bank.

2.3.3 Direct Memory Access (DMA) controller

The 'C62xx/'C67xx on-chip DMA controller allows data transfers between the internal memory and (1) external memory: (2) host port and (3) external peripherals. The DMA data transfer is performed with zero overhead and is

Conflicting instructions		Priority
Load	Load	Load using DA2 occurs first
Store	Store	Store using DA1 occurs first
Load	Store	Load always occurs first

Table 2.7 Load–store priorities

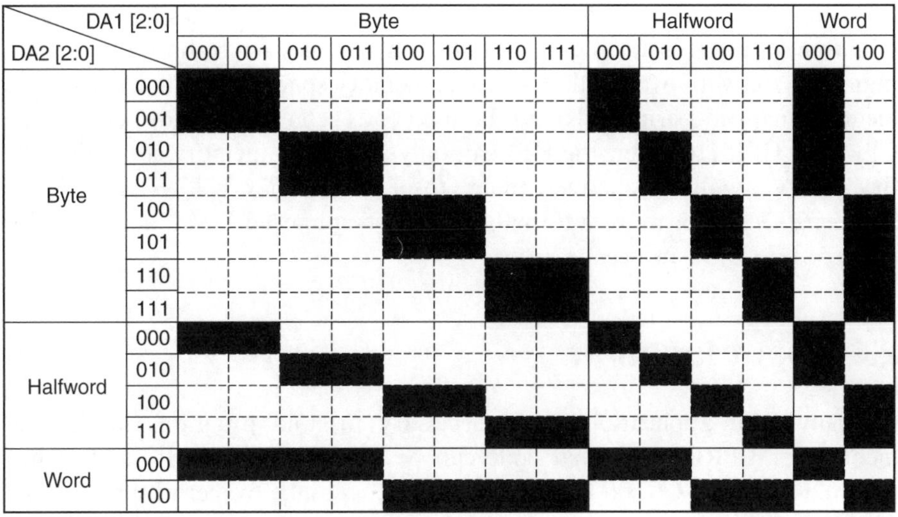

DA2 [2:0] \ DA1 [2:0]		Byte								Halfword				Word	
		000	001	010	011	100	101	110	111	000	010	100	110	000	100
Byte	000	■	■							■				■	
	001	■	■							■				■	
	010			■	■						■			■	
	011			■	■						■			■	
	100					■	■					■			■
	101					■	■					■			■
	110							■	■				■		■
	111							■	■				■		■
Halfword	000	■	■							■				■	
	010			■	■						■			■	
	100					■	■					■			■
	110							■	■				■		■
Word	000	■	■	■	■					■	■			■	
	100					■	■	■	■			■	■		■

Figure 2.12 Possible memory conflicts

transparent to the CPU, which means that the DMA and the CPU operations can be independent. Of course, if the DMA and the CPU both try to access the same memory location, arbitration will be performed by the program memory controller. The 'C62xx and the 'C67xx each have four DMA channels and one auxiliary channel dedicated for the Host Port Interface (HPI) (refer to data sheets for up-to-date figures). Detailed information and applications are given in Chapters 8 and 9.

2.4 Serial ports

The 'C62xx/'C67xx have two Multichannel Buffered Serial Ports (McBSP) allowing inexpensive interface for industry standard and user configurable serial communications with external peripherals. The serial ports support full-duplex communication and operate at a maximum speed of 40 Mbit/s per channel (refer to data sheet for up-to-date figures). Data transfer between the internal memory and the serial ports via the DMA is also supported and each serial port can handle up to 128 channels (see Chapters 3, 8 and 9 for more details).

2.5 Host Port Interface

The HPI provides a low cost interface to standard microprocessor buses. The HPI is a parallel port through which a host processor can directly access the CPU's memory. The host ports on the 'C62xx/'C67xx support 16-bit data

(refer to data sheet for up-to-date figures). However, it provides the possibility of automatically transferring 32-bit data in two successive 16-bit transfers with minimum software overhead. The whole memory space (internal, external and memory mapped peripherals) can be used for exchange of data between the CPU and HPI. However, the HPI can only access the memory space through the DMA auxiliary channel (refer to TMS320C62xx/C67xx *Peripherals Reference Guide* for more information and the enhanced DMA controller for the 'C6211/'C6711 devices).

2.6 Boot function

In many DSP applications, it is necessary to hold programs in external memory or EPROM, so that code can be loaded to the DSP device when powered-up. The 'C62xx/'C67xx offer this possibility by permitting booting from either (1) internal memory: (2) external memory or (3) HPI. These booting processes are controlled by the external pins BOOTMODE[4:0] (see TMS320C62xx/C67xx *Peripherals Reference Guide*). The three boot modes are described in the following sections.

No Boot mode

In this case the CPU starts executing code from the internal memory located at address 0.

Memory Boot mode

The 16K $\times$ 32-bit words available in external space CE1 are copied automatically to address zero by the DMA controller (see Figures 2.9a and 2.9b). The 'C62xx/'C67xx accept booting from 8-, 16- or 32-bit external memory as long as the data is ordered in little-endian format. The reorganisation of 8- and 16-bit data to 32-bit data is performed by the EMIF.

Host Boot mode

In this mode a host processor initialises the 'C6X memory space through the HPI and the external memory configuration registers. When the boot process has ended, the CPU is taken out of reset and starts executing code from address zero.

2.7 Internal timers

Two 32-bit programmable internal timers are available with the 'C62xx/'C67xx. Each timer is composed of one timer period register to host the count value and one count-up (Timer Counter) register to generate an interrupt as it reaches the value FFFFFFFFh. The count-up register is clocked at every quarter-cycle and is reloaded by the value in the timer reload register every time it reaches zero. The countdown register can be clocked internally or externally (using the TINP pin). A general-purpose pin (TOUT) is also available for signalling the rate of the internal timer to external devices (see Chapter 3).

2.8 'C62xx/'C67xx interrupts

As with most microprocessors, the 'C62xx and 'C67xx allow normal program flow to be interrupted. In response to the interruption, the CPU finishes executing the current instruction(s) and branches to a procedure which services the interrupt. To service an interrupt, the user must save the contents of the registers and the context of the current process, then service the interrupt task, restore the registers and the context of the process, and finally resume the original process (see Figure 2.13). The interrupt can come from an external device, an internal peripheral or simply a special instruction in the program.

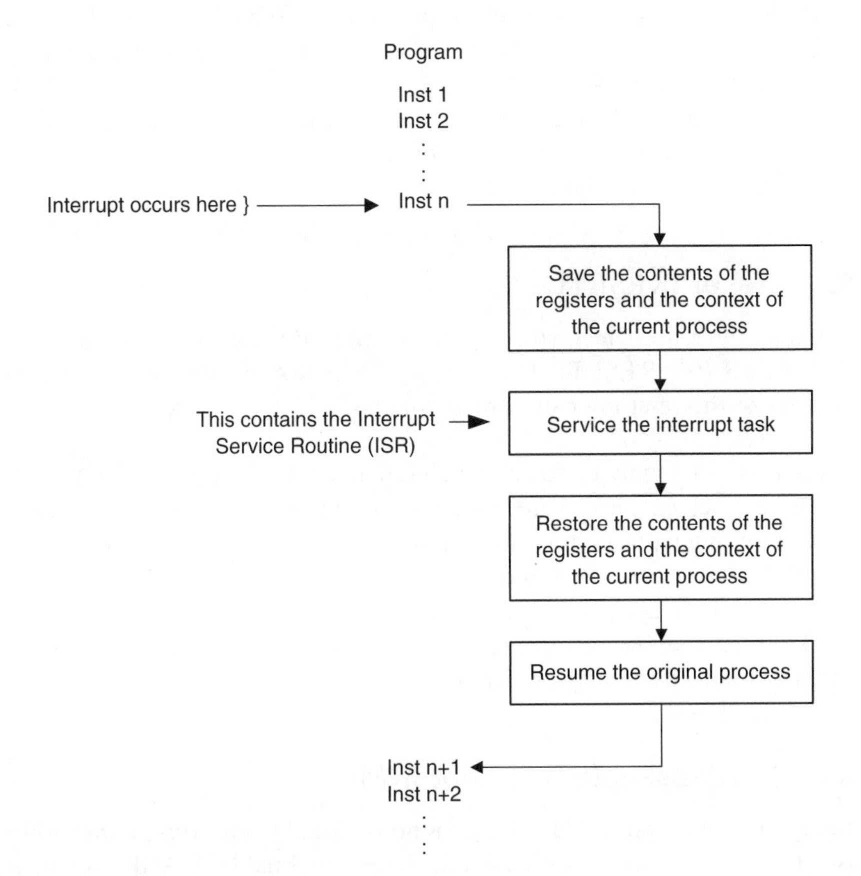

Figure 2.13 Interrupt response procedure

Type	Interrupt name	Priority
Non-maskable	$\overline{\text{RESET}}$	Highest
Maskable	NMI	↓
	INT4	
	INT5	
	INT6	
	INT7	
	INT8	
	INT9	
	INT10	
	INT11	
	INT12	
	INT13	
	INT14	
	INT15	Lowest

Table 2.8 Interrupt sources and priority

There are two main types of interrupts on the ’C62xx/’C67xx CPUs. These are the non-maskable interrupts (Reset and NMI) and maskable interrupts (INT4–INT15) (see Table 2.8).

Some of the maskable interrupts are used by the internal peripherals; others are used under software control and the rest are reserved. For more details refer to the appropriate data sheets.

2.8.1 Reset ($\overline{\text{RESET}}$)

Reset is an interrupt which stops the CPU and initialises all registers to their default values (pin $\overline{\text{RESET}}$). The reset interrupt has the highest priority. The properties of the reset interrupt are as follows:

(1) Reset pin is the only interrupt which is active low and must be held low for at least 10 clock cycles before it goes high in order to reset the device.
(2) The reset interrupt vector should always be located at address 0.
(3) With reset, the execution of instructions in progress is not completed as in the other interrupts, but is aborted.
(4) All registers are set to their default values (refer to appropriate data sheet).
(5) Reset is not affected by branches.

2.8.2 NonMaskable Interrupt (NMI)

Although the NMI pin is classified as a non-maskable interrupt, it can still be masked by clearing the NonMaskable Interrupt Enable (NMIE) bit in the control status register (CSR) (see Figure 2.14).

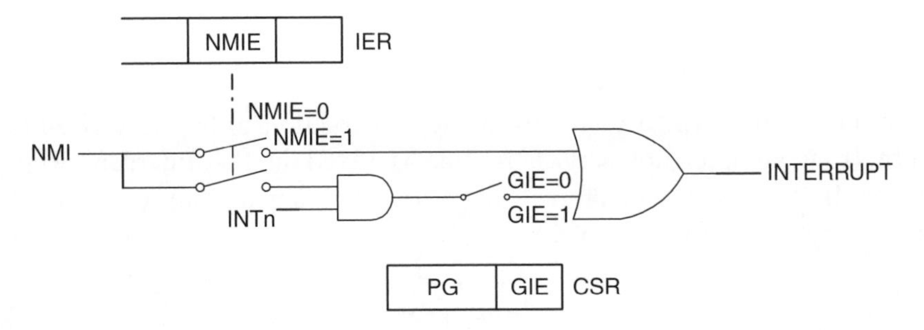

Figure 2.14 Interrupt logic

The NMIE is only set to zero upon reset and upon a non-maskable interrupt processing. This prevents interruption of an NMI by another NMI. It should be noted that the NMIE cannot be cleared by the user. If NMIE is set to zero, all interrupts (INT4–INT15) and the NMI itself are disabled.

During NMI processing, the return pointer, which contains the address of the next instruction to be executed, is stored in the NMI Return Pointer Register (NRP). Therefore, the branch (`B NRP`) instruction will cause a return to the previous program flow after servicing the NMI.

Two conditions prevent NMI from causing interrupts:

(1) The NMI Enable (NMIE) bit in the Interrupt Enable Register (IER) is set to 0.
(2) The CPU is in the delay slots of a branch (whether the branch is taken or not).

Note: The NMIE bit cannot be cleared by the user.

2.8.3 Maskable interrupts (INT4–INT15)

There are 12 maskable interrupts associated with the 'C62xx/'C67xx CPUs. These interrupts can be generated by:

- External devices (such as A/D and D/A)
- On-chip peripherals (such as internal timer)
- Software traps.

In order for a maskable interrupt to be serviced, the following conditions should be met:

(1) The corresponding interrupt occurs.
(2) The NMIE bit is set to 1.
(3) The Global Interrupt Enable (GIE) bit in the Control Status Register (CSR) is set to 1.
(4) The corresponding Interrupt Enable (IE) bit in the IER is set to 1.

(5) There is no higher priority interrupt pending.
(6) The CPU is not in the delay slots of a branch.

The return pointer which points to the next instruction to be executed after a return from an interrupt service routine is saved in the Interrupt Return Pointer (IRP) register. Therefore, return from the interrupt subroutine can be achieved by `B IRP`, for example:

```
B    .S2   IRP   ; return and moves PGIE to GIE
NOP  5           ; delay slots
```

2.8.3.1 Interrupt Service Table (IST) and Interrupt Service Table Pointer (ISTP)

When an interrupt occurs, the CPU automatically recognises the source of the interrupt and jumps to the interrupt vector location. In this location, a program is found which instructs the processor on the action(s) to be taken.

Each vector location can accommodate eight instructions which correspond to a fetch packet. Such a location is known as the Interrupt Service Fetch Packet (ISFP) address. Table 2.9 shows the interrupt sources and associated ISFP address.

Note: Each vector address is aligned on a fetch packet boundary.

When the CPU branches to the ISFP address, different possibilities exist for the Interrupt Service Routine (ISR):

- The ISR can be fitted in a single fetch packet (see Figure 2.15).
- The ISR can be fitted in multiple successive fetch packets (see Figure 2.16).

Interrupt source	ISFP address
Reset	0x0000
NMI	0x0020
Reserved	0x0040
Reserved	0x0060
INT4	0x0080
INT5	0x00A0
INT6	0x00C0
INT7	0x00E0
INT8	0x0100
INT9	0x0120
INT10	0x0140
INT11	0x0160
INT12	0x0180
INT13	0x01A0
INT14	0x01C0
INT15	0x01E0

Table 2.9 Interrupt service table (IST)

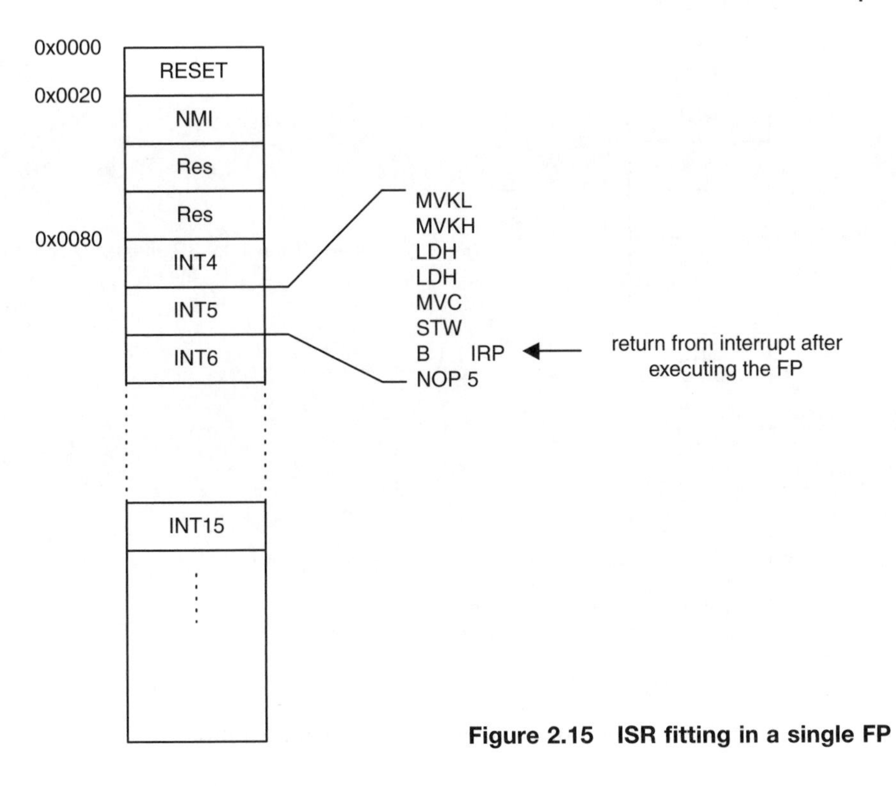

Figure 2.15 ISR fitting in a single FP

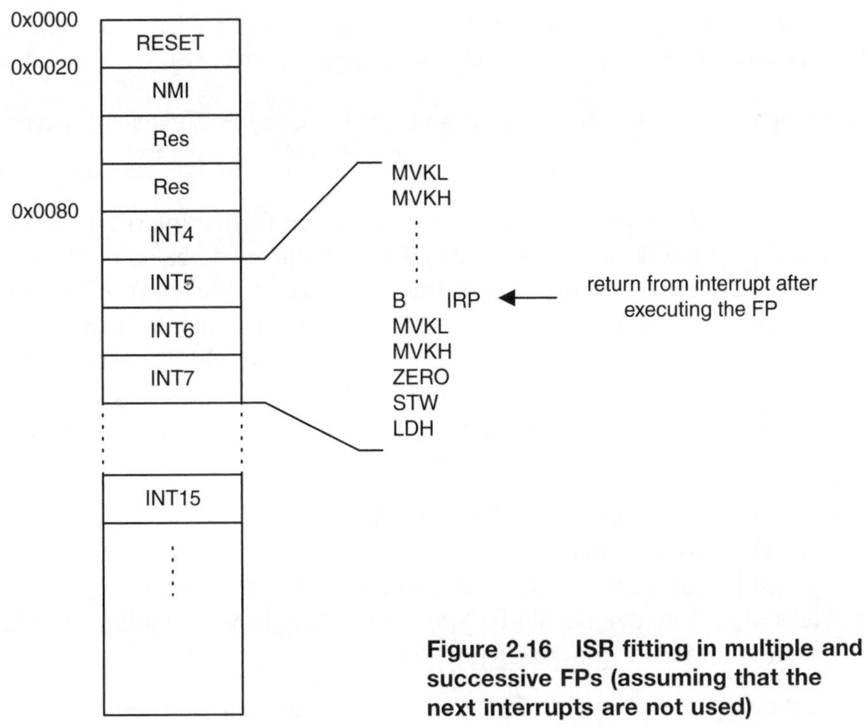

Figure 2.16 ISR fitting in multiple and successive FPs (assuming that the next interrupts are not used)

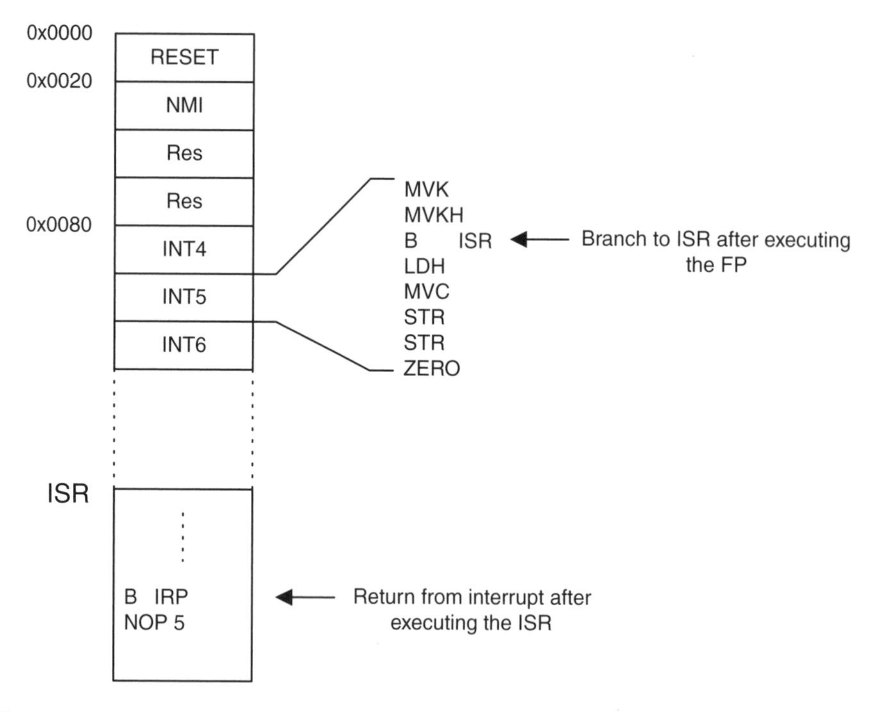

Figure 2.17 ISR is situated outside the interrupt vector table

- The ISR can be fitted in multiple non-successive fetch packets and each fetch packet can point to another fetch packet (see Figure 2.17).

> Note: If a `B ISR` does not branch to the beginning of a FP, instructions above the ISR will be ignored even if they are within the FP and in parallel.

It is the task of the programmer to write the code for the interrupt vector. If the length of the ISR can fit in one or a few FPs, there will be no need to write the ISR outside the interrupt vector (see Figures 2.15 and 2.16). If the length of the ISR does not fit into the interrupt vector, then some of the instructions and a branch to the ISR can be in the interrupt vector, and the rest of the ISR put in an appropriate program memory location as shown in Figure 2.17.

In order to create an interrupt vector, the following points should be observed:

- Code should be aligned on FP boundaries.
- The ISFP should contain a maximum of eight instructions.
- The ISFP must contain a branch to either IRP, NRP or ISR.
- The section 'vectors' should be linked to the appropriate address, typically linked to address zero (see Chapter 3).

The following sequence of code shows how to create an interrupt vector.

```
.sect "vectors"

Reset MVKL      .S2  ISR, B0
      MVKH      .S2  ISR, B0
      B         .S2  B0
      NOP
      NOP
      NOP
      NOP
      NOP

NMI   MVKL      .S2  NMI_ISR, B0
      MVKH      .S2  NMI_ISR, B0
      B         .S2  B0
      NOP
      NOP
      NOP
      NOP
      NOP

      .space (size in bytes)

INT4  B         .S2  ISR4
      NOP
      NOP
      NOP
      NOP
      NOP
      NOP
      NOP
       :
       :
       :
      .end
```

Note: In the Reset and NMI ISFP, the address of the ISR is first loaded in B0 then a branch using B0 is used, whereas in INT4 ISFP a direct branch to ISR4 is used. Both methods are valid, but a branch using a register has a higher dynamic range (32-bit) than a branch to a label having only a 21-bit signed dynamic range.

2.8.3.2 Relocating the vector table

In general, the vector table or the section 'vectors' is linked to address zero. However, in many applications, there is a need to change the location of the vector table. This is due to many factors such as:

- moving interrupt vectors to fast memory
- having multiple vector tables for use by different tasks
- boot ROM already contained in memory starting at 0x0000 address
- memory starting at location zero is external and hence there will be a need to move the vector table to internal memory to avoid bus conflict in shared memory system.

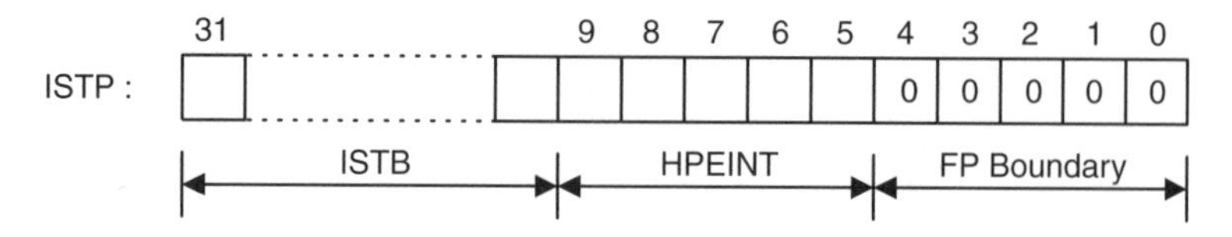

Figure 2.18 Interrupt Service Table Pointer (ISTP) register

In order to be able to relocate the vector table, the Interrupt Service Table Pointer (ISTP) register should be set up first. The ISTP is described below.

2.8.3.3 Interrupt Service Table Pointer (ISTP) register

The ISTP register allows dynamic computation of the interrupt vector, by generating the fetch packet address. This is simply achieved by concatenating the Interrupt Service Table Base (ISTB) which defines the new base address, the value of the Highest Priority Enabled Interrupt (HPEINT) and five zeroes (for bounding the ISFP on a fetch packet (FP) boundary) (see Figure 2.18).

Table 2.9 can now be generalised to include the Interrupt Service Table Base (see Table 2.10).

> Note: On Reset the ISTB is always set to zero and therefore the ISFP of reset is not reallocatable.

2.8.3.4 Interrupt Enable Register (IER)

All interrupts except the reset can be enabled or disabled (masked) individually by either setting or clearing the appropriate bits of the IER (see

Interrupt source	ISFP address
Reset	0x0000
NMI	ISTB + 0x0020
Reserved	ISTB + 0x0040
Reserved	ISTB + 0x0060
INT4	ISTB + 0x0080
INT5	ISTB + 0x00A0
INT6	ISTB + 0x00C0
INT7	ISTB + 0x00E0
INT8	ISTB + 0x0100
INT9	ISTB + 0x0120
INT10	ISTB + 0x0140
INT11	ISTB + 0x0160
INT12	ISTB + 0x0180
INT13	ISTB + 0x01A0
INT14	ISTB + 0x01C0
INT15	ISTB + 0x01E0

Table 2.10 Interrupt service table (IST) including ISTB

Figure 2.19 Interrupt Enable Register (IER)

Figure 2.19). If an interrupt occurs and the appropriate bit in the IER is set to 1 and the GIE is set, then the CPU services the interrupt (assuming that the NMIE = 1). Otherwise, if the IER is cleared to zero, no action is taken.

The following code sequence shows how to enable, for instance, interrupt 7 (INT7) and interrupt 4 (INT4) and disable all the others (the reset is always enabled).

```
MVK  0091h, B0 ;B0 = 0000 0000 0000 0000 0000 0000 1001 0001
MVC  B0, IER   ;IER is a control register and therefore can only be written to by MVC
```

Note: Reset is not maskable and writing a zero to bit 0 of the IER will have no effect.

As seen above, the non-maskable interrupt, NMI, can also be masked by using the IER. This is very useful in preventing any interruption of the processor initialisation after a reset. However, after a reset, the NMIE must be set to enable the NMI and INT4–INT15 (INT4–INT15 can only be enabled if NMIE = 1 and GIE = 1 and the appropriate IER bit(s) are set) (see Figure 2.14).

Note: At Reset, the NMIE = 0 to prevent any interrupt being serviced.

2.8.3.5 Interrupt Flag Register (IFR)

When a valid interrupt is recognised by the CPU, an appropriate bit in the Interrupt Flag Register (IFR) will be set to 1. This bit remains set until the interrupt is serviced or is cleared by the user (see Figure 2.20).

Note: The IFR can be read, but not directly written to.

To read the interrupt flag register (IFR), use:

```
MVC   IFR, B0
```

Figure 2.20 Interrupt Flag Register (IFR)

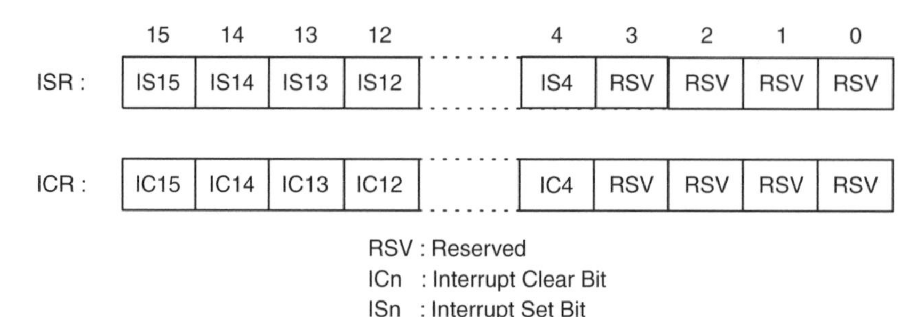

Figure 2.21 Interrupt Set Register (ISR) and Interrupt Clear Register (ICR)

2.8.3.6 Interrupt Set Register (ISR) and Interrupt Clear Register (ICR)

Access to the IFR for modification needs to be done through the Interrupt Set Register (ISR) for setting a bit and through the Interrupt Clear Register (ICR) for clearing a bit. To set or clear a bit, the corresponding bit in ISR or ICR should be set (see Figure 2.21), while writing zero to any bit has no effect. The interrupt flag bits corresponding to the non-maskable interrupt (reset and NMI) can be read but not modified, and subsequently their locations in ICR and ISR are reserved.

The following sequence of code gives an example of how to set bit 8 of the ISR and clear bit 4 of the ICR:

```
MVK   0100h, B0     ; set bit 8 to 1
MVC   B0, ISR       ; setting the pending INT8
MVK   010h, A0      ; Set bit 4 to 1
MVC   A0, ICR       ; Clear the pending INT4
```

Note: All the control registers (including ISR and ICR) can only be accessed by the MVC instruction.

2.8.3.7 Global Interrupt Enable (GIE)

The GIE bit located in bit zero of the Control Status Register (CSR) (see Figure 2.22) globally enables (GIE = 1) or disables (GIE = 0) all maskable interrupts.

When the GIE is modified, its previous value is saved in the 'Previous GIE' (PGIE). When an interrupt occurs, the CPU automatically clears the GIE to prevent any nested interrupt, and upon return from interrupt using `B IRP`, the

	31	1	0
CSR :		PGIE	GIE

Figure 2.22 Control Status Register (CSR)

PGIE is copied back to the GIE. In order to use the nested interrupt, the user must enable the GIE bit within the Interrupt Service Routine.

Note: Writing to GIE takes two cycles.

2.8.3.8 Interrupt signalling and acknowledgement (INUMx and IACK)

There are four pins (INUM0–INUM3) available to indicate to the external hardware which interrupt is being processed. For instance, if INT5 is being processed, the signals 0, 1, 0, 1 will be available on INUM3, INUM2, INUM1 and INUM0 respectively. There is also a pin (IACK) that indicates that an interrupt has occurred and is being processed. This is useful for CPU and external device handshaking.

2.9 Instruction set

The 'C6x instruction set supports arithmetic and logical operations, program control, load/store and bit management operations as well as some other operations as shown in Table 2.11. Due to the fact that different instructions are executed by different execution units, the instructions have been grouped by execution units for quick reference when programming. This is shown in Table 2.11b. Refer to the manuals for the additional instructions used by the 'C67xx processors.

Arithmetic	Logical	Program control	Load/store	Bit management	Other
ABS	AND	B	LD	CLR	IDLE
ADD	CMPEQ		MVKL	EXT	MV
ADDA	CMPGT		MVKH	LMBD	MVC
ADDK	CMPLT		ST	NORM	NOP
ADD2	NOT		STP	SET	ZERO
MPY	NEG				
MPYH	OR				
MPYHL	SHL				
MPYLH	SHR				
SMPY	SSHL				
SMPYH	XOR				
SADD					
SAT					
SSUB					
SUB					
SUBA					
SUBC					
SUB2					

Table 2.11 (a) TMS320C62xx instruction set grouped by operation types

.S unit		.L unit		.M unit	.D unit	No unit
ADD	MVKH	ABS	NEG	MPY	**ADD**	NOP
ADDK	NOT	**ADD**	OR	MPYH	ADDA	IDLE
ADD2	NEG	AND	SADD	MPYHL	LD ST	
AND	OR	CMPEQ	SAT	MPYLH	**SUB**	
B	SET	CMPGT	SSUB	SMPY	SUBA	
B IRP	SHL	CMPLT	**SUB**	SMPYH		
B NRP	SHR	LMBD	SUBC			
B reg	SSHL	MV	XOR			
CLR	STP	NORM	ZERO			
EXT	**SUB**	NOT				
MV	SUB2					
MVC	XOR					
MVKL	ZERO					

Table 2.11 *(cont.)* **(b) TMS320C62xx instruction set grouped by functional units**

Note: The ADD and SUB instructions can execute in .S, .L and .D units.

2.9.1 Writing Assembly code

The instruction syntax for the assembler is as follows:

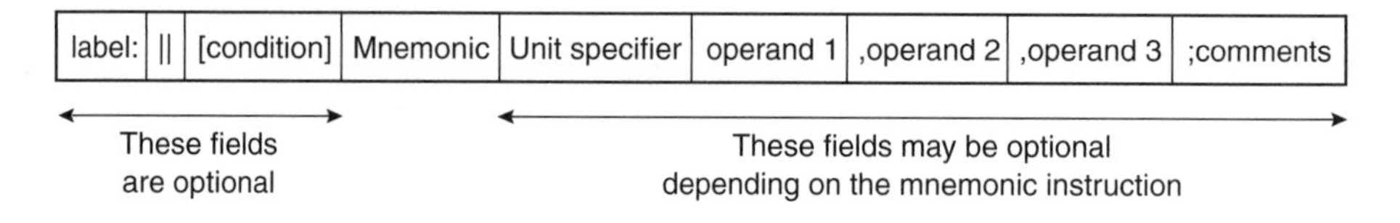

Program 2.1 shows how to write an assembly code.

```
* You can put comments when a line is started by an asterisk as shown in this line
Loop  LDH  .D1   *A8++,A2   ; Load the samples pointed to by A8 then increment the pointer
||    LDH  .D2   *B8++,B2
[B0]  SUB  .L2   B0,1,B0    ; Subtract if B0 = 0.
[B0]  B    .S1   Loop       ; Branch to loop if B0 ≠ 0.
      NOP        2          ; You can use [!B0] to branch if B0 = 0
      MPY  .M1x  A2,B2,A4
      NOP
      ADD  .L1   A4,A5,A6
```

Program 2.1 Example of assembly language syntax

Label

If a label is used it must begin in column 1; however, if it is not used then *a blank space must be left*. The colon ':' is optional.

The pipe symbol '||' indicates that the instruction which is in the same line as the pipe symbol is in parallel with the previous instruction (in Program 2.1 the pipe symbol means that the instructions (`LDH .D1 *A8++,A2`) and (`LDH .D2 *B8++,B2`) are in parallel).

Mnemonic

The mnemonic field can contain the following:

(1) Instruction (e.g. `ADD`; `MPY`; `LDH`, etc.)
(2) Directive (e.g. `.sect`; `.data`, etc.). Note that all directives start with a dot.

Unit specifier

The unit specifier field contains the name of the functional unit used for the instruction. It is optional (see Chapter 4) and therefore its field can be left blank. However, by specifying the units, the programmer has more control.

Operand fields

The fields for operand 1, operand 2 and operand 3 can contain one of the following:

(1) A register (e.g. `A1`) or a register pointer (`*A1`)
(2) A symbol (e.g. `loop`) or constant (e.g. `190`)
(3) An expression (e.g. `if label >= 100`).

Comment field

This field contains useful comments and is ignored by the assembler. The comments are always started by a semi-colon '`;`'.

Chapter 3

Software development tools and TMS320C6201 EVM overview

3.1 Introduction

To develop software for real-time applications, many issues need to be considered. The programmer must generate a source code for the application, decide which target hardware configuration to use and then use the development tools to produce an executable code to be downloaded onto the target or burned onto an EPROM. It is essential for a DSP programmer to fully understand the use and capabilities of the tools. This chapter is divided into three main parts: the first part describes the development tools, the second part describes the evaluation module (EVM), and, finally, the third part describes the codec and the use of interrupts and provides some useful programs for testing the EVM.

3.2 Software development tools

The software development tools consist of the following modules: the C compiler, assembler, linker, simulator and the code converter (see Figure 3.1). If the source code is written in C language, the code should be compiled using the Optimizing C Compiler provided by Texas Instruments (SPRU187). This compiler will translate the C source code into an assembly code. The assembly code generated by either the programmer, the compiler or the linear assembler (see Chapter 4) is then passed through the assembler that translates the code into object code. The resultant object files, any library code and a command file are all combined by the linker which produces a single executable file. The

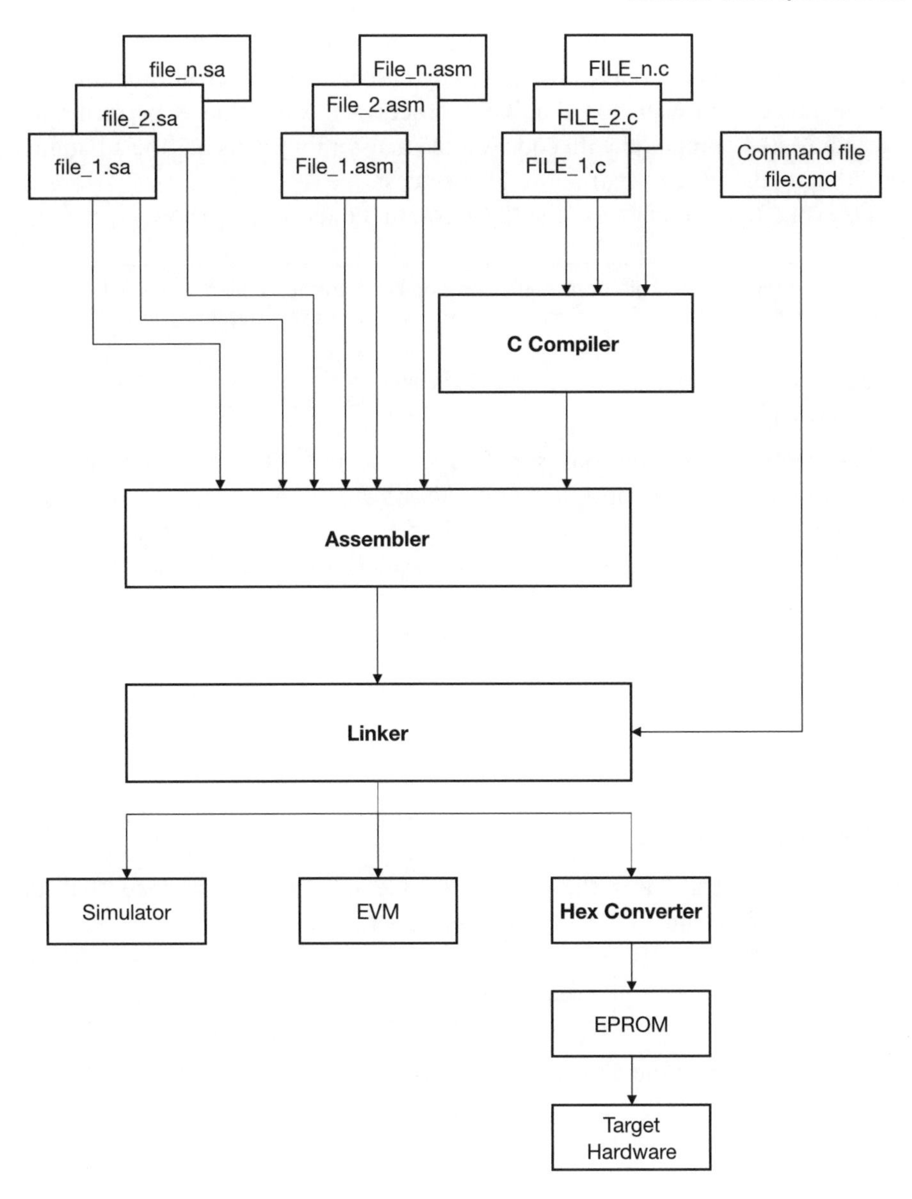

Figure 3.1 Development tools

command file mainly provides the target hardware information to the linker and is described in Section 3.2.3.1.

3.2.1 Compiler

The C code is not an executable code and therefore needs to be translated to a language that the DSP processor understands. In general, programs are written in C because of its portability, ease of use and popularity. Although for time-critical applications the assembly language is the most efficient language, the optimising C compiler for the 'C6x processors can achieve performances exceeding 70% compared with code written in assembly. This has the advantage of reducing the time-to-market and hence cost.

To evoke the compiler use the **CL6x** command as shown below:

```
CL6x FIR1.c (This command line compiles the file called FIR1.c)
```

Note: The CL6x command is not case sensitive.

The compiler uses options supplied by the user. These options provide information about the program and the system to the compiler. The most common options are described in Chapter 4. However, for a complete description of the compiler options, the reader is referred to the *Optimizing C Compiler User's Guide* (SPRU187).

The options shown in Table 3.1 can be inserted between the CL6x command and file name as shown below:

```
CL6x  -gk   FIR1.c
```

3.2.2 Assembler

The assembler translates the assembly code into an object code that the processor can execute. To evoke the assembler, type:

```
asm6x  FIR1.asm    FIR.obj
```

The above command line assembles the FIR1.asm file and generates the FIR.obj file. If the 'FIR.obj' is omitted the assembler automatically generates

Option	Description
-mv6701	Tells the compiler that the code is for the 'C6701 processor.
-k	Do not delete the assembly file (*.asm) created by the compiler.
-g	Symbolic debugging directives are generated in order to enable debugging.
-i	Specifies the directory where the #include files reside.
-s	Interlists C and assembly source statements.
-z	Adding the -z option to the command line will evoke the assembler and the linker.

Table 3.1 Common compiler options

Option	Description
-l	Generates an assembly listing file.
-s	Puts labels in the symbolic table in order to be used by the debugger.
-x	Generates a symbolic cross-reference table in the listing file (using the -ax option automatically evokes the -l option).

Table 3.2 Common assembler options

an object file with the same name as the input file but with the .obj extension, in this case FIR1.obj.

Note: The asm6x command is not case sensitive.

The assembler, as with the compiler, also has a number of 'switches' that the programmer can supply. The most common options are shown in Table 3.2.

The following command line assembles the FIR1.asm file and generates an object file called firl.obj and a listing file called firl_lst.lst.

```
asm6x  -g FIR1.asm  firl.obj  -l firl_lst.lst
```

Note: The file names are case sensitive.

3.2.3 Linker

The various object files which constitute an application are all combined by the linker to produce a single executable file. The linker also takes as inputs the library files and the command file that describes the hardware. To evoke the linker, type:

```
lnk6x  FIR1.obj  comd.cmd
```

The above command line links the FIR1.obj file with the file(s) contained in the command file comd.cmd. The linker options can also be contained in the command file.

The linker also has different options that are specified by the programmer. The most common options are shown in Table 3.3.

The following command line links the file FIR1.obj with the file(s) specified in the comd.cmd file and generates a map file (FIR1.map) and an output file (FIR1.out).

Note: The -m FIR1.map and the -o FIR1.out command could be included in the command file.

Option	Description
-o	Names an output file.
-c	Uses auto-initialisation at runtime.
-l	Specifies a library file.
-m	Produces a map file.

Table 3.3 Frequently used options for the linker

```
lnk6x FIR1.obj  comd.cmd  -m FIR1.map  -o FIR1.out
```

Note: The `lnk6x` command is not case sensitive and if `-o FIR1.out` is omitted then an `A.out` file will be generated instead.

3.2.3.1 Linker command file

The command file serves three main objectives. The first objective is to describe to the linker the memory map of the system to be used, and this is

```
/* linker command file used with examples using MAP 1 */
FIR1.obj          /* This specifies the input file */
-o FIR1.out                 /* This specifies output file */
-m FIR1.map                 /* This specifies the map file */
-l rts6201.lib              /* This specifies the library file */

MEMORY
{
 INT_PROG_MEM (RX)   :origin = 0x00000000 length = 0x00010000
 SBSRAM_PROG_MEM (RX)  : origin = 0x00400000 length = 0x00020000
 SBSRAM_DATA_MEM (RW)  : origin = 0x00420000 length = 0x00020000
 SDRAM0_DATA_MEM (RW)  : origin = 0x02000000 length = 0x00400000
 SDRAM1_DATA_MEM (RW)  : origin = 0x03000000 length = 0x00400000
 INT_DATA_MEM (RW)     : origin = 0x80000000 length = 0x00010000
}

SECTIONS
{
 .vec:    load = 0x00000000
 .text:   load = SBSRAM_PROG_MEM
 .const:  load = INT_DATA_MEM
 .bss:    load = INT_DATA_MEM
 .data:   load = INT_DATA_MEM
 .cinit   load = INT_DATA_MEM
 .pinit   load = INT_DATA_MEM
 .stack   load = INT_DATA_MEM
 .far     load = INT_DATA_MEM
 .sysmem           load = SDRAM0_DATA_MEM
 .cio     load = INT_DATA_MEM
}
```

Figure 3.2 Command file for the TMS320C6201 EVM (`comd.cmd`)

specified by 'MEMORY {...}'. The second objective is to tell the linker how to bind each section of the program to a specific section as defined by the 'MEMORY' area, which is specified in the 'SECTIONS {...}'. The third objective is to supply the linker with the input and output files, and options of the linker. An example of a command file for the TMS320C6201 evaluation module (EVM) is shown in Figure 3.2.

As with all embedded systems, the command file is indispensable for real-time application. The linker options specified in the CL6x command can be specified within the command file as shown in Figure 3.2.

3.2.4 Compile, assemble and link

The command CL6x combined with the -z linker option can accomplish the compiling, assembling and linking stages all with a single command as shown below:

```
CL6x  -gs FIR1.c  -z comd.cmd
```

3.3 TMS320C6201 Evaluation Module, EVM

The EVM is a relatively low-cost demonstration board. It allows one to evaluate the performance of the 'C6201 processors. The EVM is designed to be used either as a plug-in-card using a PCI expansion slot inside a PC, or as a stand-alone unit controlled by an XDS510 (for PCs) or XDS510WS (for SUN stations) emulator. In the latter case the EVM will require an external power supply. The basic layout of the EVM is shown in Figure 3.3.

3.3.1 EVM features

The EVM's features are as follows:

- Uses the TMS320C6201 DSP processor
- DSP clock: 33.25 MHz, 50 MHz, 133 MHz or 200 MHz
- External memory:
  - 64K × 32, 133 MHz SBSRAM
  - 1M × 32, 100 MHz SDRAM (bank 0)
  - 1M × 32, 100 MHz SDRAM (bank 1)
- Expansion memory interface provided
- PCI or JTAG Emulator Interface
- Stereo codec interface (44.1 kHz sampling)
- Code generation tools and support library included.

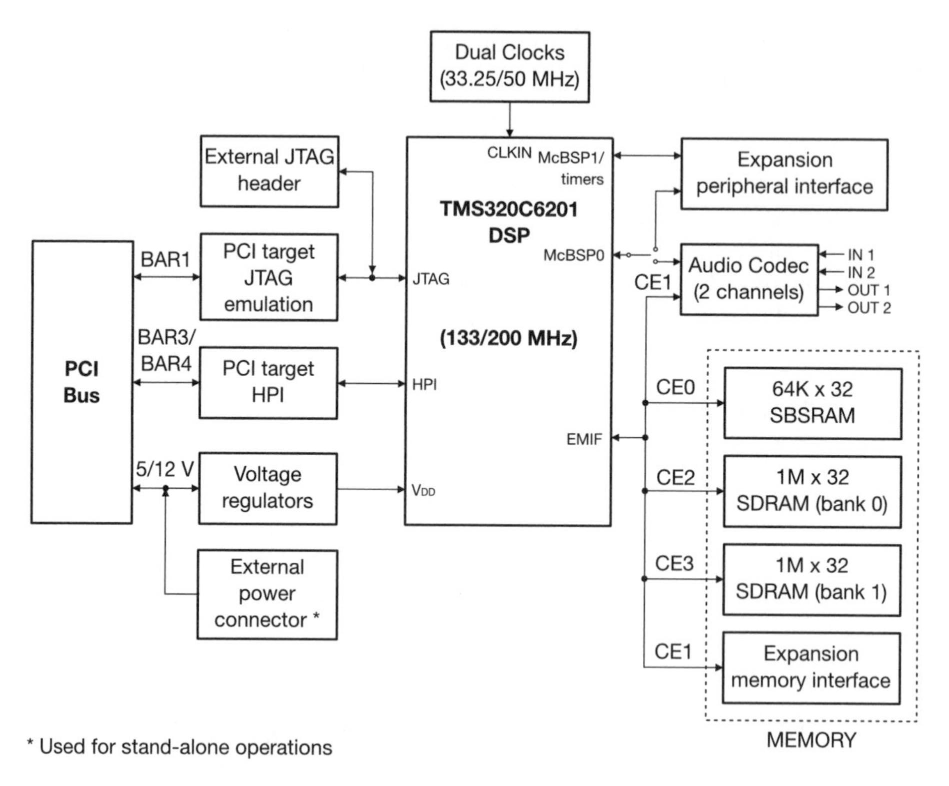

Figure 3.3 EVM layout

3.3.2 Using interrupts

The 'C62xx/'C67xx have 16 interrupt sources each (see Table 3.4). However, the CPU can handle only 12 interrupts at a time; these are INT4–INT15. Therefore it is not possible to utilise all the 16 interrupt sources at the same time. To overcome this problem the user has the possibility of selecting any 12 interrupt sources and then mapping them onto any of the 12 CPU interrupts. This is achieved by using two interrupt multiplexers (see Figure 3.4). For

31 30	29 26	25	24 21	20	19 16	15 14	13 10	9	8 5	4	3 0
reserved	INTSEL9	reserved	INTSEL8	reserved	INTSEL7	reserved	INTSEL6	reserved	INTSEL5	reserved	INTSEL4

(a) Low register of the interrupt multiplexer (memory mapped at address 0x019C0004)

31 30	29 26	25	24 21	20	19 16	15 14	13 10	9	8 5	4	3 0
reserved	INTSEL15	reserved	INTSEL14	reserved	INTSEL13	reserved	INTSEL12	reserved	INTSEL11	reserved	INTSEL10

(b) High register of the interrupt multiplexer (memory mapped at address 0x019C0000)

Figure 3.4 Interrupt multiplexer registers: (a) low register; (b) high register

Interrupt source	Interrupt acronym	Selection number (binary)
Host processor to DSP interrupt	DSPINT	0000
Timer 0 interrupt	TINT0	0001
Timer 1 interrupt	TINT1	0010
EMIF SDRAM timer interrupt	SD_INT	0011
External interrupt pin 4	EXT_INT4	0100
External interrupt pin 5	EXT_INT5	0101
External interrupt pin 6	EXT_INT6	0110
External interrupt pin 7	EXT_INT7	0111
DMA channel 0 interrupt	DMA_INT0	1000
DMA channel 1 interrupt	DMA_INT1	1001
DMA channel 2 interrupt	DMA_INT2	1010
DMA channel 3 interrupt	DMA_INT3	1011
McBSP 0 transmit interrupt	XINT0	1100
McBSP 0 receive interrupt	RINT0	1101
McBSP 1 transmit interrupt	XINT1	1110
McBSP 1 receive interrupt	RINT1	1111

Table 3.4 Interrupt sources available on the TMS320C6x processors

instance, if the external interrupt 4 (EXT_INT4) is to be mapped to the CPU interrupt 5, then the INTSEL5 field needs to be set to 4. The default setting is shown in Table 3.5.

Mapping an interrupt source to a CPU interrupt can be done using assembly or C code. The following sequence of code shows how to map the source interrupt 5 to the CPU interrupt 4.

In Assembly:

```
MVKL  0x019c0004,A0  ; Load the 16 lsb address of the interrupt mutiplexer into A0
MVKH  0x019c0004,A0  ; Load the 16 msb address of the interrupt mutiplexer into A0
MVKL  00101b,A5      ; Load the interrupt vector number of EXT_INT5 to A5
STB   A5,*A0         ; Store A5 into INTSEL4.
```

CPU interrupt (related INTSEL *field*)		Interrupt source (INTSEL *value*)		Description	
INT4	(INTSEL4)	EXT_INT4	(0100)	EXT_INT4	mapped to INT4
INT5	(INTSEL5)	EXT_INT5	(0101)	EXT_INT5	mapped to INT5
INT6	(INTSEL6)	EXT_INT6	(0110)	EXT_INT6	mapped to INT6
INT7	(INTSEL7)	EXT_INT7	(0111)	EXT_INT7	mapped to INT7
INT8	(INTSEL8)	DMA_INT0	(1000)	DMA_INT0	mapped to INT8
INT9	(INTSEL9)	DMA_INT1	(1001)	DMA_INT1	mapped to INT9
INT10	(INTSEL10)	SD_INT	(0011)	SD_INT	mapped to INT10
INT11	(INTSEL11)	DMA_INT2	(1010)	DMA_INT2	mapped to INT11
INT12	(INTSEL12)	DMA_INT3	(0011)	DMA_INT3	mapped to INT12
INT13	(INTSEL13)	DSPINT	(0000)	DSPINT	mapped to INT13
INT14	(INTSEL14)	TINT0	(0001)	TINT0	mapped to INT14
INT15	(INTSEL15)	TINT1	(0010)	TINT1	mapped to INT15

Table 3.5 Default mapping

In C language: see Program 3.1.

```
/* Program showing how to set an interrupt in C */
# include <intr.h>

void Timer1_ISR (void);        /* ISRs do not exchange arguments. */
INTR_MAP_RESET();              /* TINT mapped to INT15 by default */
intr_hook (Timer1_ISR, CPU_INT15);
INTR_ENABLE (CPU_INT15);       // Enable cpu interrupt line 15
INTR_GLOBAL_ENABLE();          // Enable global interrupt.

interrupt void Timer1_ISR (void)
{
/* Include your code here */

/* each time the internal timer generates an interrupt, this function is called */
}
```

Program 3.1 Interrupt setting in C language

Note: The interrupt directive tells the compiler that the function used is an interrupt service routine and therefore all the relevant registers are to be saved before entering the ISR and restored after exiting it.

There are two API (Application Programming Interface) functions for mapping the source interrupts to the CPU interrupts. These are `intr_map (int cpu_intr, int isr)` and `INT_MAP_RESET()`. Refer to *Peripheral Support Library* (SPRU273) for a complete description.

3.3.2.1 Linking an Interrupt Service Routine (ISR) to the interrupt vector

It has been shown previously how to map a source interrupt to a CPU interrupt. However, this alone is not sufficient because the end result would be to evoke the interrupt for servicing an ISR as shown in Chapter 2. Therefore the address of the ISR has to be in the vector interrupt table. To achieve this a library function **`intr_hook()`** is provided (SPRU273). The ISR functions written in C code do not exchange parameters and also need to be declared with the **`interrupt`** directive as shown in Program 3.1. Figure 3.5 illustrates the overall operation of the interrupts.

3.3.3 Stereo codec overview

The EVM includes a 16-bit stereo codec (CS4231A) from Crystal (www.cirrus.com). The codec supports full-duplex transfer and can be interfaced to other DSP chips or an ASIC. The codec has a programmable sampling frequency which also depends on the crystal frequency selected, as shown in Table 3.6. The sampling frequency ranges from 5.5 kHz to 48 kHz.

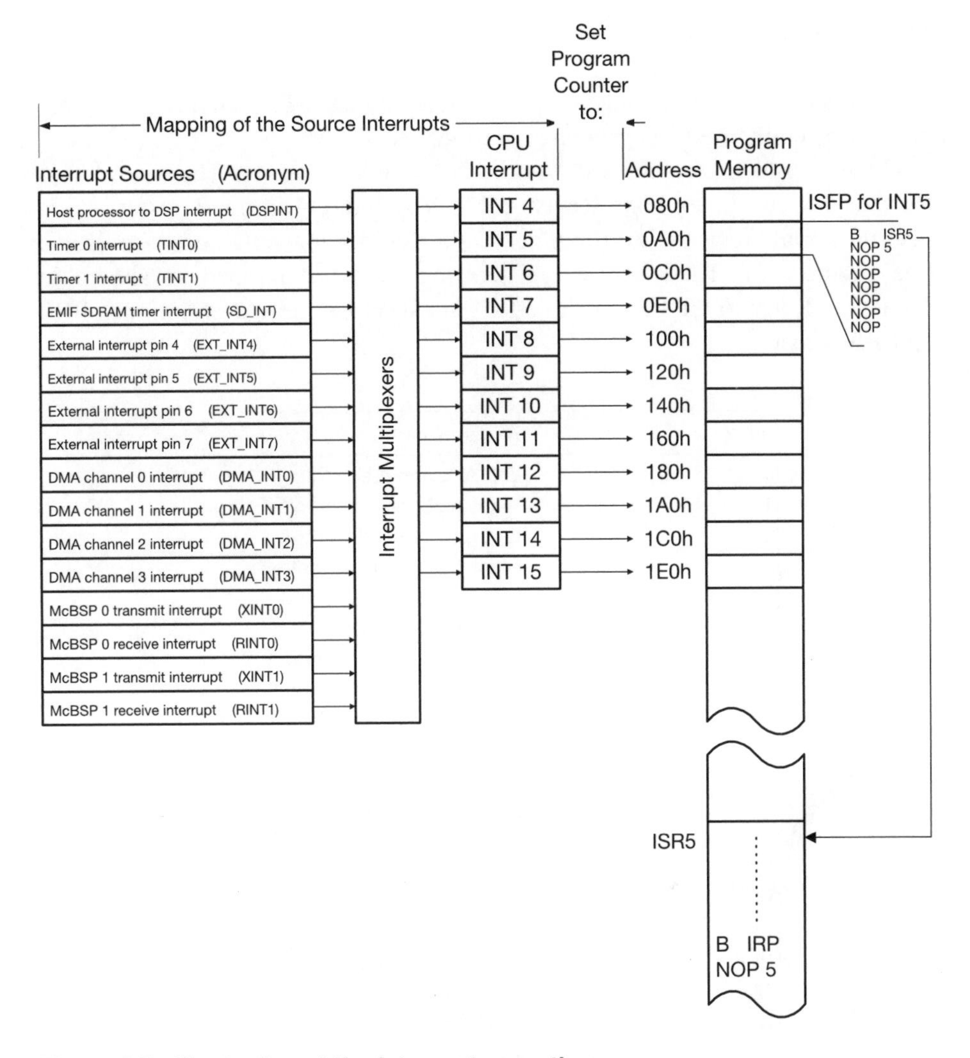

Figure 3.5 Illustration of the interrupt operations

Sampling frequency (kHz) XTAL1 (24.567 MHz)	Sampling frequency (kHz) XTAL2 (16.9344 MHz)
8.0	5.51
9.0	6.62
16.0	11.025
27.42	18.9
32.0	22.05
48.0	33.075
–	37.8
–	44.1

Table 3.6 Sampling frequency range of the stereo codec

The crystals used (XTAL1 and XTAL2) have frequencies of 24.567 MHz and 16.944 MHz respectively.

The CS4231A codec is fully programmable and has 32 internal registers that need to be configured for proper operation: see Table 3.7.

These registers are accessed by two 8-bit index registers (Index Address Register R0 and Index Data Register R1). The Index Address Register selects the appropriate register and the Index Data Register loads the register selected (see Figure 3.6). These codec registers are memory mapped as shown in Table 3.8. Refer to the CS4231A codec data sheet for more descriptions of R2 and R3 registers.

Index	Register name
I0	Left ADC Input control
I1	Right ADC Input Control
I2	Left Aux #1 Input Control
I3	Right Aux #1 Input Control
I4	Left Aux #2 Input Control
I5	Right Aux #2 Input Control
I6	Left DAC Output Control
I7	Right DAC Output Control
I8	Fs & Playback Data Format
I9	Interface Configuration
I10	Pin Control
I11	Error Status and Initialisation
I12	MODE and ID (MODE2 bit)
I13	Loopback Control
I14	Playback Upper Base Count
I15	Playback Lower Base count
I16	Alternate Feature Enable I
I17	Alternate Feature Enable II
I18	Left Line Input Control
I19	Right line Input Control
I20	Timer Low Base
I21	Timer High Base
I22	RESERVED
I23	Alternate Feature enable III
I24	Alternate Feature status
I25	Version / Chip ID
I26	Mono Input & Output Control
I27	RESERVED
I28	Capture DATA Format
I29	RESERVED
I30	Capture Upper Count Base Count
I31	Capture Lower Base Count

Table 3.7 Index Register and associated registers

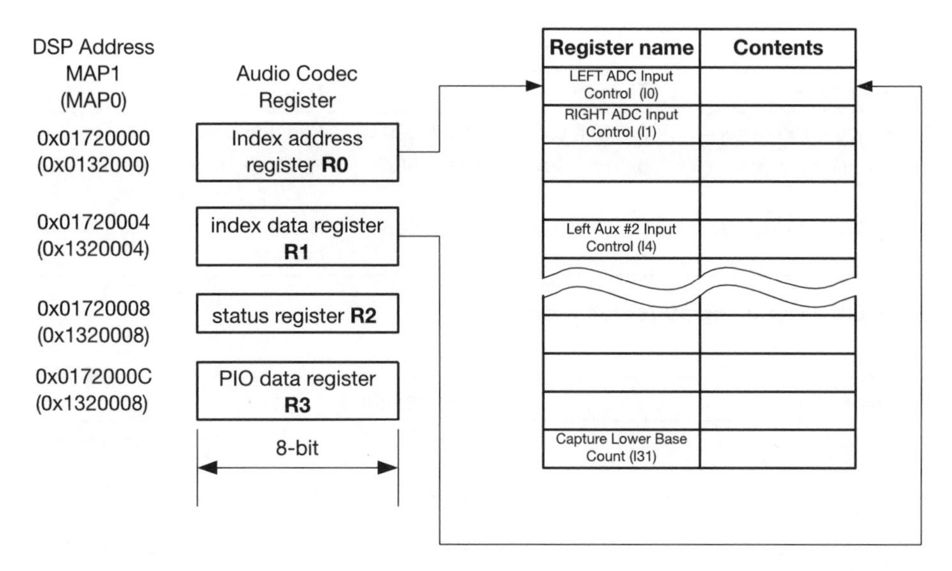

Figure 3.6 Illustration of the codec registers' settings

Register name	DSP address MAP1 (MAP0)	Access
Index address register (R0)	0x0172000 (0x1320000)	Read/write
Index data register (R1)	0x0172004 (0x1320004)	Read/write
Status register (R2)	0x0172008 (0x1320008)	Read/write
PIO data register (R3)	0x017200c (0x132000C)	Read/write

Table 3.8 Codec registers' map

3.3.4 Configuring and using the internal timers

The two internal timers are controlled by six memory-mapped registers (three registers for each timer). These registers are:

- The **Timer Control Registers** which set the operating modes of the timers.
- The **Timer Period Registers** which hold the number of the timer input clock cycles to count.
- The **Timer Counters** which hold the current value of the incrementing counter (see Table 3.9). The timer clock cycle is the CPU clock cycle divided by four.

To configure the operation of the internal timers, the Timer Control Registers (TCR) have to be set correctly (refer to the 'C62xx/'C67xx *Peripherals Reference Guide*, SPRU190). In the default mode, all the bits of the TCR are set to zero.

The peripheral support library (SPRU273) supplied with the EVM module contains 10 functions for manipulating the two timers. These functions are listed in Table 3.10.

Register name	Address		Description
	Timer 0	Timer 1	
Timer Control Register	0x01940000	0x01980000	Sets the operating mode.
Timer Period Register	0x01940004	0x01980004	Holds the number of the timer clock cycles to count.
Timer Counter Register	0x01940008	0x01980008	Holds the current value of the counter.

Table 3.9 Timer registers' locations

Function	Description
`TIMER_AVAILABLE(chan)`	Check for the availability of the internal timers
`TIMER_RESET(chan)`	Reset the timer registers
`TIMER_CLK_EXTERNAL(chan)`	Select the external clock as the source
`TIMER_CLK_INTERNAL(chan)`	Select the internal clock as the source
`TIMER_INIT(chan, ctrl, per)`	Set the timer period and control registers
`TIMER_SET_PERIOD(chan, val)`	Set the timer period register
`TIMER_SET_COUNT(chan, val)`	Set the timer count register
`TIMER_START(chan)`	Start the timer
`TIMER_STOP(chan)`	Stop the timer
`TIMER_RESUME(chan)`	Resume the operation of the timer

Table 3.10 Timers' library functions

3.3.4.1 Programming the internal timers in C language

Program 3.2 shows how to program the internal timers using the functions and macros supplied by the peripheral support library. In this example the timer 1 rate, f_s, is set to 48 kHz. Equation [3.1] can be used to find the number n, to be loaded into Period Register 1.

$$f_s = \frac{\frac{F_{\text{clk}}}{4}}{n} \qquad [3.1]$$

F_{clk} is the CPU clock cycle and is 133 MHz (the EVM is clocked at 133 MHz), therefore:

$$n = \frac{F_{\text{clk}}}{4 \times f_s} = \frac{133,000,000}{4 \times 48,000} = 692.7 \cong 693$$

3.3.5 Testing the EVM

Now that we have seen how to use interrupts, the stereo codecs and the internal timer, let us write a program in C language to test the functionality of the software and hardware involved. In this program a signal is sampled at

```
void set_up_timer1 (void)

{
int period;
float rate;
/* Set the timer 1 to the default configuration, see page 9-4 SPRU190B */
TIMER_RESET(1);

/* Check if the timer 1 is available, if so then do the following initialisation */
if (TIMER_AVAILABLE (1))
  {

  TIMER_INIT(1,0x0200,693,0xffff);
/* 1:      select the timer 1 */
/* 0x0200: set the clksrc bit to 1, so that the timer 1 input clock source is the CPU clock/4 */
/* 693:    is the value of the timer period register */
/* 0xffff: is the value of the timer counter */

  TIMER_SET_PERIOD(1,693);           /* This function sets the period register */
/* this is not required since it is done with the function TIMER_INIT() */

/* This function reads the Timer period from the period register */
  period = TIMER_GET_PERIOD(1);

  rate = 1/((period *4) / 133000000);  /* rate of the timer 1 */

  printf ("\n timer1 rate is % .2f  Hz \n", rate);
  }
/* To stop and start the timer 1 use the following functions: */
/* TIMER_START(1);  This starts the timer 1 */
/* TIMER_STOP(1);   This stops the timer 1  */

}
```

Program 3.2 C routine for setting up the timers

44.100 kHz by the codec and is sent back as shown in Figure 3.7. The transfer function of the codec given in Figure 3.8 shows that the magnitude is flat between about 4 Hz and 19 kHz. The DC to 5 Hz signals are attenuated by a decoupling capacitor on the EVM. The program is self-explanatory and is shown in Program 3.3. In this example the sampling rate is generated by the codec using the function '`codec_change_sampling_rate()`'. To compile and

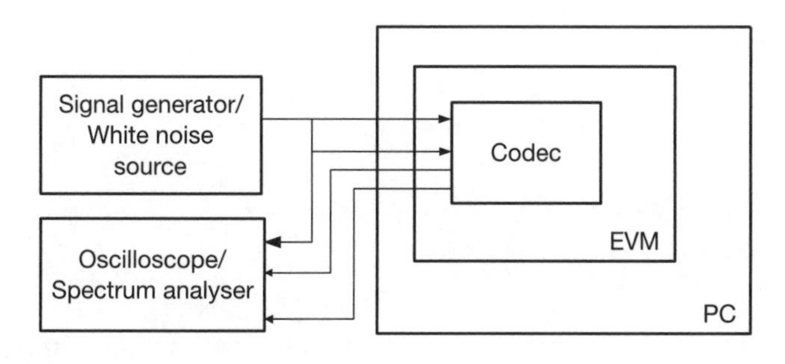

Figure 3.7 Testing set-up

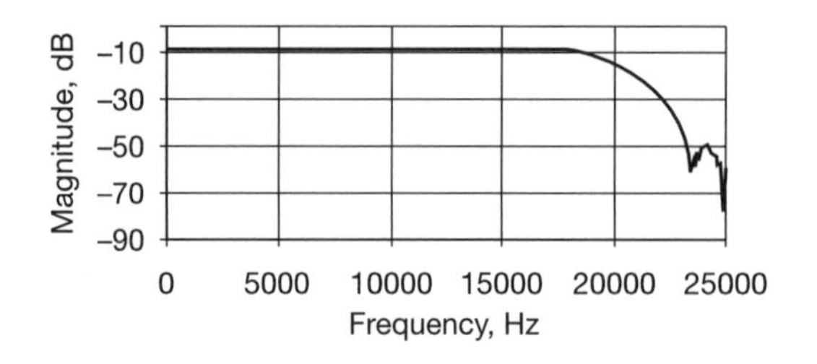

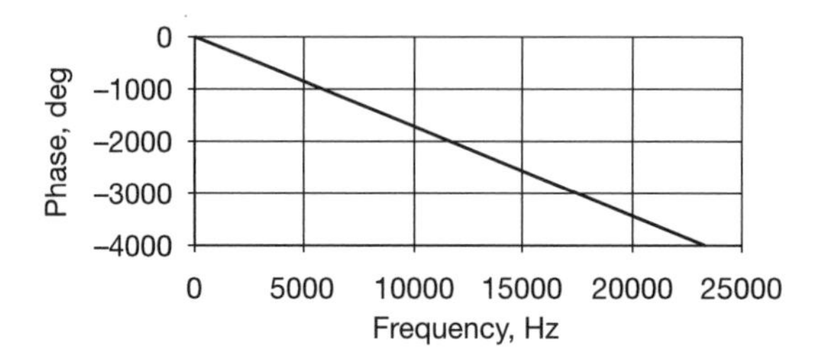

Figure 3.8 Transfer function of the codec when using a sampling rate of 44.100 kHz

link this file, a batch file has been created (see `inout.bat`); all the input and output files can be found in the `F:\DSPCODE\Inout (Chap3)\` directory.

```
/*-------------------------------------------------------------------
        File:         inout.c
        Description:  Samples signal from line-input and redirect to
                      line-out. Uses on-board codec and mcbsp0.
--------------------------------------------------------------------*/

#include <stdio.h>
#include <string.h>
#include <common.h>
#include <board.h>
#include <stdlib.h>
#include <mcbspdrv.h>
#include <intr.h>
#include <regs.h>

#define SAMPLING_RATE 44100

extern int inicodec (int sampling_rate);
interrupt void isr_rint0 (void)
{
  int sample;
  sample = MCBSP0_DRR;
  MCBSP0_DXR = sample;
  return;
}
```

Program 3.3 EVM6X test programs: (a) main program, `inout.c`

Program 3.3 continued

```
void Init_Interrupts ()
{
  intr_reset();                     // Reset the interrupt system,
                                    // disable all interrupts.
  INTR_CLR_FLAG (CPU_INT15);        // Clear previous interrupt request
  INTR_ENABLE (CPU_INT15);          // Enable cpu interrupt line 15
  intr_map(CPU_INT15,ISN_RINT0);    // Link receive interrupt of serial
                                    // port 0 to interrupt line 15.
  intr_hook (isr_rint0,CPU_INT15);  // Assign the interrupt service routine
}

void main(void)
{
  evm_init();                       // Initialise EVM board.
  Inicodec (SAMPLING_RATE);         // Initialise codec, adjust sampling rate.
  Init_Interrupts ();               // Initialise interrupts and hook isr.
  INTR_GLOBAL_ENABLE();             // Enable global interrupt.
  for (;;);                         // Main loop, does nothing.
}
```

```
/*----------------------------------------------------------------------
        File:         inicodec.c
        Description: Setups the on-board codec to sample from line-in,
                     at 16-bit 2-channel per sample
                     Set up mcbsp0 to 32-bit per sample.
----------------------------------------------------------------------*/

#include <common.h>
#include <codec.h>
#include <mcbspdrv.h>

int inicodec (int sampling_rate)
{
  Mcbsp_dev dev;
  Mcbsp_config mcbspConfig;
  mcbsp_drv_init();
  dev = mcbsp_open(0);
  if (dev == 0) return FALSE;
  memset(&mcbspConfig,0,sizeof(mcbspConfig));
  mcbspConfig.loopback = FALSE;
  mcbspConfig.tx.update = TRUE;
  mcbspConfig.tx.clock_polarity = CLKX_POL_RISING;
  mcbspConfig.tx.frame_sync_polarity = FSYNC_POL_HIGH;
  mcbspConfig.tx.clock_mode = CLK_MODE_EXT;
  mcbspConfig.tx.frame_sync_mode = FSYNC_MODE_EXT;
  mcbspConfig.tx.phase_mode = SINGLE_PHASE;
  mcbspConfig.tx.frame_length1 = 0;
  mcbspConfig.tx.word_length1 = WORD_LENGTH_32;
  mcbspConfig.tx.frame_ignore = FRAME_IGNORE;
  mcbspConfig.tx.data_delay = DATA_DELAY0;
```

Program 3.3 EVM6X test programs: (b) codec initialisation, `inicodec.c`

Program 3.3 continued

```
    mcbspConfig.rx.update = TRUE;
    mcbspConfig.rx.clock_polarity = CLKX_POL_RISING;
    mcbspConfig.rx.frame_sync_polarity = FSYNC_POL_HIGH;
    mcbspConfig.rx.clock_mode = CLK_MODE_EXT;
    mcbspConfig.rx.frame_sync_mode = FSYNC_MODE_EXT;
    mcbspConfig.rx.phase_mode = SINGLE_PHASE;
    mcbspConfig.rx.frame_length1 = 0;
    mcbspConfig.rx.word_length1 = WORD_LENGTH_32;
    mcbspConfig.rx.frame_ignore = FRAME_IGNORE;
    mcbspConfig.rx.data_delay = DATA_DELAY0;
    if (mcbsp_config(dev,&mcbspConfig)==ERROR) return FALSE;

    MCBSP_ENABLE (0, 3);           // Bring Serial Port 0 out of reset, 3 means both rx and tx.

    codec_init ();                 // Initialise on-board codec, enabling it as well.
    codec_serial_port_enable (SERIAL_32BIT);             // Enable codec serial i/o and set
                                                         // serial format to 32-bit.
    codec_change_sample_rate (sampling_rate, TRUE);      // Set the sampling rate.
    codec_adc_control (LEFT, 0.0, FALSE, LINE_SEL);      // Unmute left ADC, set gain select
                                                         // source from line input
    codec_adc_control (RIGHT, 0.0, FALSE, LINE_SEL);     // Unmute right ADC, set gain select
                                                         // source from line input
    codec_line_in_control (LEFT, MAX_AUX_LINE_GAIN, TRUE);  // Set left line input volume and
                                                         // mute direct loopback to DAC
    codec_line_in_control (RIGHT, MAX_AUX_LINE_GAIN, TRUE); // Set right line input volume and
                                                         // mute direct loopback to DAC
    codec_dac_control (LEFT, 0.0, FALSE);                // Unmute left DAC and set volume.
    codec_dac_control (RIGHT, 0.0, FALSE);               // Unmute right DAC and set volume.
    return TRUE;
}
```

Chapter 4

Software optimisation

4.1 Introduction

Software optimisation is the process of manipulating software code to achieve two main goals: faster execution and smaller code size.

To implement efficient software, the programmer must be familiar with the processor architecture, the language(s) (i.e. C and assembly) used, the compiler, assembler and linear assembler features and the code that they generate. The preferred and most supported high level language used for programming DSP processors is the ANSI C language. In fact most DSP manufacturers support only the ANSI C language.

Code written in C is in general not processor-specific and is portable. However, code written in assembly is processor-specific and not portable, but runs faster than code written in C and consumes less memory.

4.1.1 Code optimisation procedure

Figure 4.1 illustrates the procedure for optimising software code. In the first step the developer must make sure that the algorithm to be implemented is fully functional and 'optimised'. In the second step, the algorithm can be implemented in ANSI C language without any optimisation. If the code is operational and the execution speed is adequate, then there is no need to develop the code further; however, if the code is functional but the execution time is not satisfactory, then the code will need to be further optimised. If all optimisations supported by the compiler still do not produce a satisfactory result, the developer needs to progress to the third step, which uses the linear assembler. In general only some functions need to be implemented in linear assembly; these functions can be determined by the profiler. If the

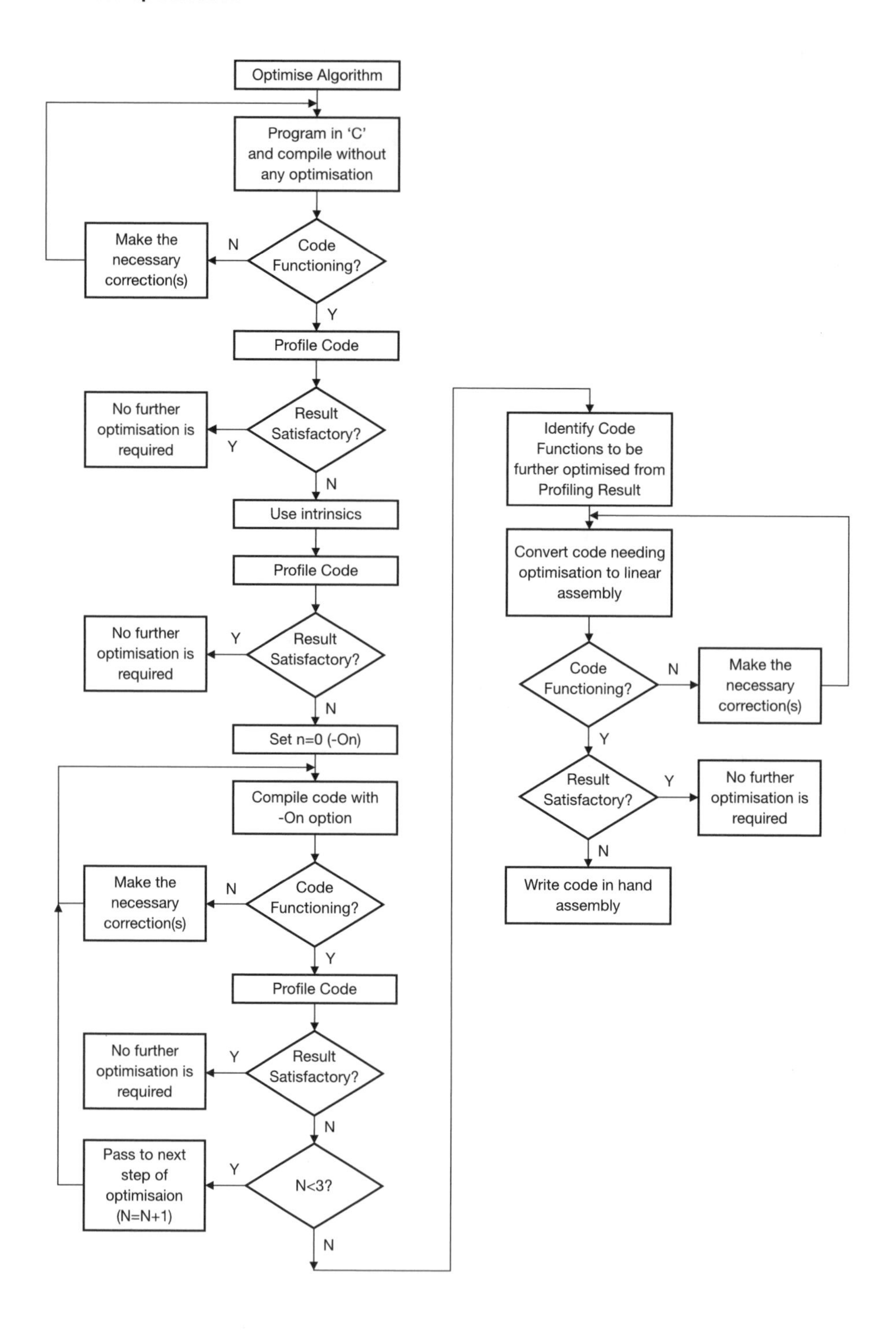

Figure 4.1 Optimisation flow procedure

Duty Free Shopping – the Simple Way

Duty Free is available after the first meal service. Explore the Inflight Gift Gallery catalogue in the seatback pocket and complete this order form. The flight attendants will collect the form during the duty free sales.

Name: ____________________

Seat # ____________________

Credit Card ☐ | Cash ☐

Page #	Quantity	Item Name	US $
______	______	____________________	______
______	______	____________________	______
______	______	____________________	______
______	______	____________________	______

Total: $ ______

FORM #6178 3/00

results obtained are still not satisfactory, the developer can move to the final stage and code the critical part of the algorithm in hand-scheduled assembly language.

4.1.2 The C compiler options

The 'C6x optimising C compiler uses the ANSI C source code and can perform optimisation currently up to about 80% compared with a hand-scheduled assembly. However, to achieve this level of optimisation, knowledge of different levels of optimisation is essential. Optimisation is performed at different stages and levels as shown in Figure 4.2 and Table 4.1. Figure 4.2 shows the different stages of the compiler passes. The C code is first passed through the Parser, which mainly performs pre-processing functions and generates an intermediate file (`.if` file). In the second stage the file produced by the parser is supplied to the optimiser, which performs most of the optimisation and produces the `.opt` file. In the third and last stage, the `.opt` file is supplied to the code generator which always performs several additional optimisations regardless of whether

-o0	• Performs control-flow-graph simplification • Allocates variables to registers • Performs loop rotation • Eliminates unused code • Simplifies expressions and statements • Expands calls to functions declared inline
-o1	• **Performs all -o0 optimisations, plus:** • Performs local copy/constant propagation • Removes unused assignments • Eliminates local common expressions
-o2	• **Performs all -o1 optimisations, plus:** • Performs software pipelining • Performs loop optimisations • Eliminates global common sub-expressions • Eliminates global unused assignments • Converts array references in loops to incremented pointer form • Performs loop unrolling • The optimiser uses -o2 as the default if you use -o without an optimisation level.
-o3	• **Performs all -o2 optimisations, plus:** • Removes all functions that are never called • Simplifies functions with return values that are never used • Inlines calls to small functions • Reorders function declarations so that the attributes of called functions are known when the caller is optimised • Propagates arguments into function bodies when all calls pass the same value in the same argument position • Identifies file-level variable characteristics

Table 4.1 Optimisation levels of the optimising compiler

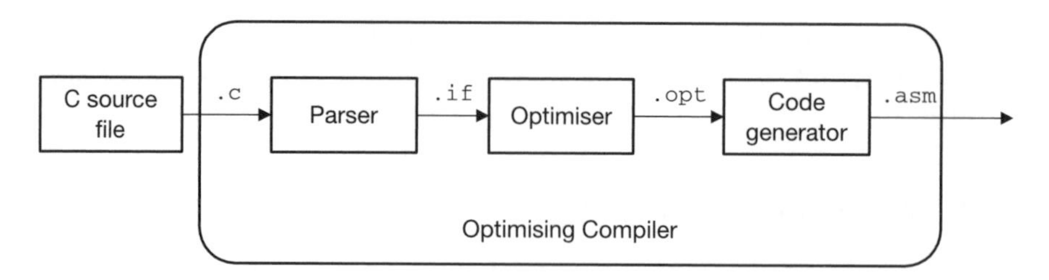

Figure 4.2 Illustration of the different stages of the optimising compiler

the optimiser is selected or not. (Refer to the TMS320C6x *Optimizing C Compiler User's Guide*, SPRU187, for a complete description.)

In order to have further control over the Parser or Optimiser, different options are available as shown in Tables 4.2a and 4.2b.

Option	Description
`-pf`	Generates function prototype listing file
`-pk`	Allows K&R compatibility
`-pl`	Generates preprocessed listing
`-pm`	Combines source files to perform program-level optimisation

Table 4.2a Most common options that control the Parser

Option	Description
`-o0`	Optimises register usage
`-o1`	Uses `-o0` optimisations and optimises locally
`-o2`	or `-o` Uses `-o1` optimisations and optimises globally
`-o3`	Uses `-o2` optimisations and optimises file

Table 4.2b Most common options that control the Optimiser

4.2 Assembly optimisation

To develop an appreciation of how to optimise code, let us optimise an FIR filter algorithm, which is represented by Equation [4.1]:

$$y[n] = \sum_{k=1}^{N} h[k]x[n-k] \qquad [4.1]$$

For simplicity we can rewrite Equation [4.1] by assuming that we can reorder samples at each sampling instant. This will lead to Equation [4.2].

$$y[n] = \sum_{i=1}^{N} h[i]x[i] \qquad [4.2]$$

To implement Equation [4.2] we need to perform the following steps:

(1) Load the samples $x[i]$
(2) Load the coefficients $h[i]$
(3) Multiply $x[i]$ and $h[i]$
(4) Add $(x[i].h[i])$ to the current content of the accumulator
(5) Repeat steps (1) to (4) $N - 1$ times
(6) Store the value in the accumulator to y.

These can be interpreted in 'C6x code as set out in Program 4.1.

```
       MVKL  .S2   0,B0       ; Initialise the loop counter
       MVKL  .S1   0,A5       ; Initialise the accumulator
loop   LDH   .D1   *A8++,A2   ; Load the samples x[i]
       LDH   .D1   *A9++,A3   ; Load the coefficients h[i]
       NOP         4          ; Add "nop 4" because the LDH has a latency of 5.
       MPY   .M1   A2,A3,A4   ; Multiply x[i] and h[i]
       NOP                    ; Multiply has a latency of 2 cycles
       ADD   .L1   A4,A5,A5   ; Add "x [i].h[i]" to the accumulator
[B0]   SUB   .L2   B0,1,B0    ; )
[B0]   B     .S1   loop       ; } loop overhead
       NOP         5          ; ) The branch has a latency of 6 cycles
```

Program 4.1 Assembly code for implementing an FIR filter

If we represent the flow of instructions on a cycle-by-cycle basis as shown in Table 4.3, we can see that for each cycle, at most one of the units is active and therefore the code as it is written is not optimised. It is clear from Table 4.3 that in order to optimise the code we need to:

(1) Use instructions in parallel, which means that multiple units will be operating on the same cycle.
(2) Remove the `NOP`s (put code in place of `NOP`s).
(3) Unroll the loop (see Section 4.2.3).
(4) Use word access instead of halfword access (see Section 4.2.4).

Let us now take each case separately and try to apply it to the code shown above.

4.2.1 Parallel instructions

Looking at Table 4.3 we see that the .D2 unit is unused and therefore the `LDH` instruction in cycle 2 can be moved to be executed in cycle 1 in the .D2 unit. This can be written as

```
       LDH   .D1   *A8++,A2
||     LDH   .D2   *B9++,B5   ; Notice that the registers come from the register
                              ; file B since .D2 is now used
```

Source/ cycle	.D1	.D2	.L1	.L2	.M1	.M2	.S1	.S2	NOP
1	LDH								
2	LDH								
3									NOP
4									NOP
5									NOP
6									NOP
7					MPY				
8									NOP
9			ADD						
10				SUB					
11							B		
12									NOP
13									NOP
14									NOP
15									NOP
16									NOP

Table 4.3 Iteration interval table for an FIR filter

The `SUB` instruction in cycle 10 could also be moved to cycle 9, and this can be written as

```
        ADD   .L1   A1,A2,A1
  ||    SUB   .L2   B10,1,B10
```

The other instructions cannot be put in parallel since the result of one unit is used as an input to the following unit. In general, up to eight instructions can be put in parallel and therefore to achieve the current maximum performance of 2400 MIPS for the 'C62xx (at 300 MHz), or 1 Gflop for the 'C67xx, all eight units should be used in parallel.

Note: For maximum performance, the Execute Packet (instructions to be executed in the same cycle) should contain eight instructions.

4.2.2 Removing the `NOP`s

Ten cycles have been 'wasted' using `NOP` instructions in the code in Table 4.3. To optimise the code further, the `NOP` instructions can be replaced by useful code. Since the `SUB` and the `B` (branch) instructions are independent of the rest of the code, then by rearranging some of the code, some `NOP`s can be eliminated as shown in Program 4.2.

Notice that the `ADD .L1` and `SUB .L2` are not used in parallel, since the `SUB` instruction has moved up with the branch instruction, and only three `NOP`s instead of 10 are being used.

```
loop  LDH   .D1   *A8++,A2   ; Load the samples x(i)

||    LDH   .D2   *B9++,B5
[B0]  SUB   .L2   B0,1,B0
[B0]  B     .S1   loop
      NOP         2          ; the 5 NOPs required for the branch instruction are replaced by
                               (NOP 2, MPY and NOP)
      MPY   .M1   A2,B3,A4
      NOP
      ADD   .L1   A4,A5,A5
                             ← The branch occurs here
```

Program 4.2 Optimised assembly code for implementing an FIR filter

4.2.3 Loop unrolling

The SUB and B instructions consume at least two extra cycles per iteration (this is known as the branch overhead). If instead of looping using the SUB and B instructions, we simply replicate the code unlooped, the branch overhead can be removed completely and the code can be reduced by at least two instructions per iteration. It is clear that with loop unrolling the code size has increased (see Program 4.3).

```
      LDH   .D1   *A8++,A2    ; Start of iteration 1
||    LDH   .D2   *B9++,B5
      NOP         4
      MPY   .M1X  A2,B3,A4    ; Use of cross path
      NOP
      ADD   .L1   A4,A5,A5
      LDH   .D1   *A8++,A2    ; Start of iteration 2
||    LDH   .D2   *B9++,B5
      NOP         4
      MPY   .M1   A2,B3,A4
      NOP
      ADD   .L1   A4,A5,A5
;           :
;           :
;           :
      LDH   .D1   *A8++,A2    ; Start of iteration n
||    LDH   .D2   *A8++,B5
      NOP         4
      MPY   .M1   A2,B5,A4
      NOP
      ADD   .L1   A4,A5,A5
```

Program 4.3 Unlooped code

4.2.4 Word access

The 'C62xx devices have two 32-bit data buses for data memory access and therefore two 32-bit data can be loaded into the registers at any one time. In addition to this the 'C62xx devices have variants of the multiplication instruction to support different operations (see Chapter 2: .M1 unit). Using these two features, the previous code can be rewritten as shown in Program 4.4.

```
loop
        LDW   .D1   *A8++,A2   ; 32-bit word is loaded in a single cycle
||      LDW   .D2   *B9++,B5
        NOP         4
[B0]    SUB   .L2   B0,1,B0
[B0]    B     .S1   loop
        NOP         2
        MPY   .M1   A2,B5,A4
||      MPYH  .M2   A2,B5,B4
        NOP
        ADD   .L1   A4,A5,A5
||      ADD   .L2   B4,B5,B5
        ADD   .L1X  A5,B5,A4
```

Program 4.4 Double word access

By loading words and using MPY and MPYH instructions, the execution time has been halved since in each iteration two 16-by-16-bit multiplications are performed. Care must be taken in dividing the loop count by 2, and adding the content of A3 and B3 when the loop ends.

4.2.5 Optimisation summary

It has been shown that there are four complementary methods for code optimisation. Using instructions in parallel, filling the delay slots or replacing NOPs with useful code and using the load word (LDW) instruction, increases the performance and reduces the code size. However, by using the loop unrolling method, the performance improves at the cost of a larger code size. Filling NOPs by reshuffling instructions can be a very tedious task. However, it is shown below that by using software pipelining procedures it can be simplified and optimised.

4.3 Software pipelining

The main objective of software pipelining is to optimise code associated with loops. The loop code is optimised by scheduling instructions in parallel and eliminating or replacing the NOPs with useful code. Due to the fact that multiple units are available on the 'C6x devices and also due to the fact that instructions have different latencies, code optimisation can be a complex task. However, by using the compiler options **-o2** or **-o3** as shown in Section 4.1, or by using the assembler optimiser as shown in Section 4.4, the burden of software pipelining can be left to the tools. To define the problem let us return to the FIR code as shown again in Program 4.5.

```
        LDH    .D1
||      LDH    .D2
        NOP           4
        MPY    .M1
        NOP
        ADD    .L1
;              :
;              :
;              :
        LDH    .D1
||      LDH    .D2
        NOP           4
        MPY    .M1
        NOP
        ADD    .L1
```

Program 4.5 Unoptimised assembly code

If we consider a table representing all units for 'all' cycle numbers and fill the appropriate boxes with the appropriate instructions, we can form a clear view of the resources used for each cycle (see Table 4.4). It is clear that each loop iteration takes eight cycles and at most one or two units are used. However, if we advance each loop by seven cycles, as shown in Table 4.5, the code still executes properly. From cycle 8 to cycle 10, four units are used by the code, and they execute in parallel. In this case we can say that we have a single cycle loop. As can be seen from Tables 4.5 and 4.6, the code can be split into three sections (prologue, kernel and epilogue).

As the name suggests, software pipelining is the process of putting code in a pipeline as shown in Table 4.5. Software pipelining is only concerned with loops since the repeatability of the code is exploited. It is evident from Table 4.5 that the loop kernel iterates the same code for each cycle.

4.3.1 Software pipelining procedure

To optimise code as shown above can be a very tedious task, especially when the loop code does not fit in a single cycle. To make code optimisation a simple procedure, it is suggested that:

(1) The code is written in linear assembly fashion. This provides a clear view of the algorithm. There is no need to specify the units, registers or delay slots (`NOP`s) as these will be taken care of in the last two steps.
(2) The algorithm is drawn on a dependency graph to illustrate the flow of data of the algorithm.
(3) List the resources (functional units, registers and cross paths) required to determine the minimum number of cycles required of each loop.

Unit/ cycle	.D1	.L2	.M1	.M2	.S1	.S2	.L1	.L2	NOP
(1)	LDH	LDH							
(2)									NOP
(3)									NOP
(4)									NOP
(5)									NOP
(6)			MPY						
(7)									NOP
(8)							ADD		
(9)	LDH	LDH							
(10)									NOP
(11)									NOP
(12)									NOP
(13)									NOP
(14)			MPY						
(15)									NOP
(16)							ADD		
(17)	LDH	LDH							
(18)									NOP
(19)									NOP
(20)									NOP
(21)									NOP
(22)			MPY						
(23)									NOP
(24)							ADD		
(25)	LDH	LDH							
(26)									NOP
(27)									NOP
(28)									NOP
(29)									NOP
(30)			MPY						
(31)									NOP
(32)							ADD		
(33)	LDH	LDH							
(34)									NOP
(35)									NOP
(36)									NOP
(37)									NOP
(38)			MPY						
(39)									NOP
(40)							ADD		
(41)	LDH	LDH							
(42)									NOP
(43)									NOP
(44)									NOP
(45)									NOP
(46)			MPY						
(47)									NOP
(48)							ADD		

Table 4.4 Iteration interval table for an FIR filter

	Unit/ cycle	.D1	.D2	.M1	.M2	.S1	.S2	.L1	.L2
Prologue	(1)	LDH	LDH						
	(2)	LDH	LDH						
	(3)	LDH	LDH						
	(4)	LDH	LDH						
	(5)	LDH	LDH						
	(6)	LDH	LDH	MPY					
	(7)	LDH	LDH	MPY					
Loop kernel	(8)	LDH	LDH	MPY				ADD	
	(9)	LDH	LDH	MPY				ADD	
	(10)	LDH	LDH	MPY				ADD	
Epilogue	(11)			MPY				ADD	
	(12)			MPY				ADD	
	(13)			MPY				ADD	
	(14)			MPY				ADD	
	(15)			MPY				ADD	
	(16)							ADD	
	(17)							ADD	

Table 4.5 Iteration interval table for an FIR filter

(1) Prologue	In this section the code is building up and its length is the length of the unroll loop minus one. In this case it is 7 (= 8 – 1).
(2) Loop kernel	Each execute packet in this section contains all instructions required for executing one loop.
(3) Epilogue	Contains the rest of the code necessary for completing the algorithm.

Table 4.6 Different sections of the code

(4) Create a scheduling table that shows instructions executing on the appropriate units, on a cycle-by-cycle basis. This table is drawn with the help of the dependency graph.
(5) Generate the final assembly code.

To gain experience of hand-optimisation using software pipelining, an FIR code is taken as an example and the five steps are shown below.

4.3.1.1 Writing linear assembly

```
Loop     LDH   *p_to_a,a
         LDH   *p_to_b,b
         MPY   a,b,prod
         ADD   sum,prod,sum
         SUB   count, 1, count
[count]  B     Loop
```

This code does not specify any unit or delay slots. Furthermore, all the registers are represented by symbolic names which make the code more readable.

4.3.1.2 Creating a dependency graph

Before creating the dependency graph, the algorithm first needs to be written in linear assembly language as shown above. Creating the dependency graph consists of three steps as shown below.

Step 1: Draw the nodes and paths

In this step each instruction is represented by a node and the node is represented by a circle. Outside the circle the instruction is written and inside the circle the register holding the result is written. The nodes are then connected by paths showing the data flow (conditional paths are represented by dashed lines). This is shown in Figure 4.3.

Step 2: Write the number of cycles it takes for each instruction to complete executing

The `LDH` takes five cycles, the `MPY` takes two cycles, the `ADD` and `SUB` take one cycle each and the `B` instruction takes six cycles to complete executing. The number of cycles should be written along the associated data path. This is shown in Figure 4.4.

Step 3: Assign functional units to each node

Since each node represents an instruction, it is advantageous to start allocating units to instructions which require a specific unit. In other words start by allocating units to nodes associated with load, store and branch. We do not need to be concerned with the multiply instruction, since multiplication can only be performed in .M units and no other instruction can be executed in .M units. This is shown in Figure 4.5. At this stage the units have been specified but not the side to which they will be allocated.

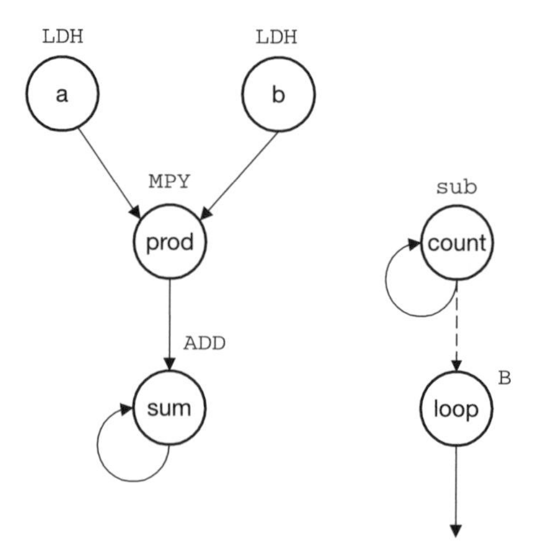

Figure 4.3 Dependency graph of an FIR filter: step 1

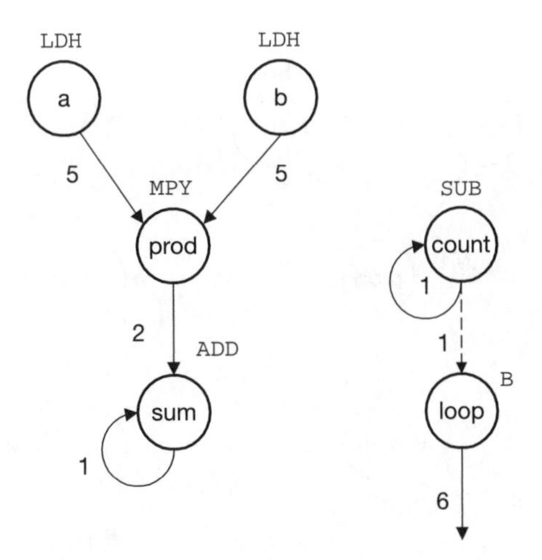

Figure 4.4 Dependency graph of an FIR filter: step 2

Step 4: Data-path partitioning

To optimise code, we need to make sure that a maximum number of units are used with a minimum of cross paths (see Chapter 2). To make this visible from the dependency graph, a dashed line is drawn on the graph to separate the two sides (see Figure 4.6).

4.3.1.3 Resource allocation

In this step all the resources are tabulated as shown in Table 4.7.

It is clear from Table 4.7 that the resources have not been exceeded, that is to say that none of the units have been reused, and also only one cross path

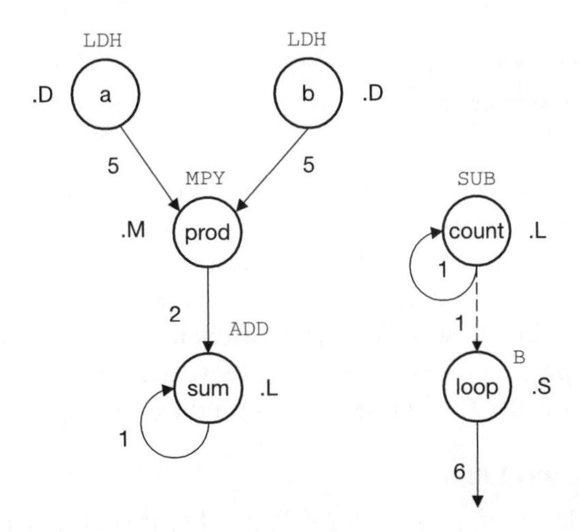

Figure 4.5 Dependency graph of an FIR filter: step 3

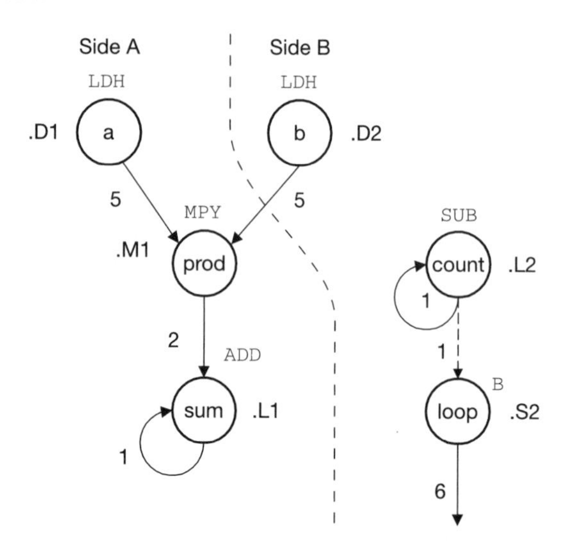

Figure 4.6 Final dependency graph of an FIR filter: step 4

Units available	Number used	Cross paths available	Number used	Register used
.L1	1	X1	1	sum
.S1	0	T1	0	
.D1	1	X2	0	a
.M1	1	T2	0	prod
.L2	1			count
.S2	1			
.D2	1			b
.M2	0			

Table 4.7 Resource allocation

and five registers have been used. From this we can conclude that the code can execute in a single cycle. Although only five registers have been named in this example, we still have to account for the registers used as address pointers; in this case we have the two addresses used by the load instructions. Although it is not necessary at this stage to link the registers used to the registers in files A and B, it is an appropriate time to do so since we are dealing with resources in this step. Finally, Table 4.8 can be used to produce the register allocation.

4.3.1.4 Scheduling table

From the dependency graph shown in Figure 4.6, it is clearer to visualise the data flow from one unit to the other. However, the picture will be more

Register file A	Symbolic registers
A0	
A1	&a
A2	a
A3	prod
A4	sum
A5	

Symbolic registers	Register file B
count	B0
&b	B1
b	B2
	B3
	B4
	B5

Table 4.8 Register allocation

complete by showing instructions executing on a cycle-by-cycle basis. This can be done by what is known as 'the scheduling table'. The table has two entries: one represents the execution units and the other represents the cycles (see Table 4.9). The number of cycles required for drawing the table is equal to the number of cycles found in the longest path of the dependency graph. This makes sense since the end of the longest path represents the end of the algorithm. In this particular example, the maximum number of cycles is eight $(= 5 + 2 + 1)$.

From Figure 4.6 we can complete Table 4.9, to generate the final scheduling table as shown in Table 4.10. We notice that on the first cycle, the two loads are executed (fill `cycle1/.D1` and `cycle1/.D2`). In order to supply the multiplication unit with the destination contents of the load instructions, the multiplication operation has to be delayed by five cycles (fill `cycle6/.M1`). Two cycles after the multiplication, the addition can be processed (fill `cycle8/.L1`).

Now that we have finished with the main part of the dependency graph, let's move on to the program control part. In this case we would like to branch at the beginning of the program as soon as the addition is performed. To do so

Cycle/units	Cycle1	Cycle2	Cycle3	Cycle4	Cycle5	Cycle6	Cycle7	Cycle8
.L1								
.L2								
.S1								
.S2								
.M1								
.M2								
.D1								
.D2								

Table 4.9 Scheduling table

	Prologue							Loop	Epilogue						
Cycle / Unit	1	2	3	4	5	6	7	8	9	10	11	12	13	14	15
.D1	LDH	LDH	LDH	LDH	LDH	LDH	LDH	LDH							
.D2	LDH	LDH	LDH	LDH	LDH	LDH	LDH	LDH							
.L1								ADD	ADD	ADD	ADD	ADD	ADD	ADD	ADD
.L2		SUB	SUB	SUB	SUB	SUB	SUB	SUB							
.S1															
.S2			B	B	B	B	B	B							
.M1						MPY	MPY	MPY	MPY	MPY	MPY	MPY	MPY		
.M2															

Table 4.10 Scheduling table

we need to schedule the branch instruction so that it executes just after the `ADD` instruction (cycle9). Since the branch instruction has a latency of six, it is scheduled in cycle 3 (fill `cycle3/.S2`).

The `SUB` instruction should occur one cycle before the branch instruction and therefore should be scheduled in cycle 2. So far we have determined the cycles in which each instruction starts to be active. From that cycle, the same instruction is repeated for all the other cycles. In cycle 8, a single-cycle loop is achieved, and hence the next cycles are identical. In practical situations the loop count is a finite number, and in order to include it in the scheduling table, we need to create the epilogue.

The epilogue can be created by removing the loop overhead (`B` and `SUB` instructions) and instructions from the prologue of the main loop code on a cycle-by-cycle basis. For example, to create the epilogue for cycle 9, we need to perform the following subtraction:

Single cycle loop	Subtract	Loop overhead	Subtract	Prologue	Result: Epilogue
LDH || LDH || MPY || ADD || SUB || B	(minus) -	SUB || B	(minus) -	LDH || LDH	MPY || ADD

4.3.1.5 Generating assembly code

From Table 4.10 we can generate the assembly code as shown in Program 4.6. Notice that the single-cycle loop can be repeated n times ($n = N - 7$), and the total number of iterations will be equal to N. This shows that the loop count is not always equal to the number of algorithm iterations.

```
; Cycle 1                  LDH   .D1    *A1++,A2
                 ||        LDH   .D2    *B1++,B2

; Cycle 2                  LDH   .D1    *A1++,A2
                 ||        LDH   .D2    *B1++,B2
                 ||        SUB   .L2    B0,1,B0

; Cycle 3- 4 and 5         LDH   .D1    *A1++,A2
                 ||        LDH   .D2    *B1++,B2
                 || [B0]   SUB   .L2    B0,1,B0
                 || [B0]   B     .S2    loop

; Cycle 6 and 7            LDH   .D1    *A1++,A2
                 ||        LDH   .D2    *B1++,B2
                 || [B0]   SUB   .L2    B0,1,B0
                 || [B0]   B     .S2    loop
                 ||        MPY   .M1x   A2,B2,A3

; Cycle 8 to N             LDH   .D1    *A1++,A2; add the 'loop' label
                 ||        LDH   .D2    *B1++,B2
                 || [B0]   SUB   .L2    B0,1,B0
                 || [B0]   B     .S2    loop
                 ||        MPY   .M1x   A2,B2,A3
                 ||        ADD   .L1    A4,A3,A4

; Cycle N + 1 to N + 5
                 ||        MPY   .M1x   A2,B2,A3
                 ||        ADD   .L1    A4,A3,A4

; Cycle N + 6 to N + 7
                           ADD   .L1    A4,A3,A4
```

Program 4.6 Code obtained from the scheduling table

4.4 Linear assembly

In the previous sections it has been shown that code optimisation for loops can be achieved by the software pipelining technique. This has been done the hard way using pen and paper. However, with the assembly optimiser, optimisation for loops can be made very simple. The tools accept code that is written in a linear fashion without taking into account the delay slots or even specifying the functional units, and by using symbolic variable names instead of registers as shown in Program 4.7.

```
        ZERO    sum
loop    LDH     *p_to_a,a
        LDH     *p_to_b,b
        MPY     a,b,prod
        ADD     sum,prod,sum
        SUB     B0,1,B0
        B       loop
```

Program 4.7 Linear assembly code representing an FIR filter

For the tools to understand which part of the code is written in linear assembly, two directives are required, the first indicating the start of the code (`.proc`) and the second indicating the end of the code (`.endproc`). The tools also require that all symbolic registers used have to be declared using the `.reg` directive as shown in Program 4.8.

```
        .proc
        .reg p_to_a, a, p_to_b, b, prod, sum
        ZERO    sum
loop    LDH     *p_to_a,a
        LDH     *p_to_b,b
        MPY     a,b,prod
        ADD     sum,prod,sum
        B       loop
        .endproc
```

Program 4.8 Linear assembly code

As with high-level languages, procedures need to exchange arguments. To achieve that, arguments are passed to the procedure through registers. These registers need to be declared at the beginning of the code and their contents moved to the appropriate symbolic registers by using the `MV` instruction. Similarly, arguments which are required to be returned to other functions must first be transferred from symbolic variables to file registers and these registers should be declared at the end of the procedure, as shown in Program 4.9, use `.cproc`/`.endproc` directives to treat the linear assembly as a c callable function (SPRU187).

```
        .proc     A4, B4, A6, B3           ; B3 not to be used (return address)
        .reg      p_to_a, a, p_to_b, b, prod
        MV        A0, p_to_a
        MV        B4, p_to_b
        MV        A6, count
        ZERO      sum
loop    LDH       *p_to_a,a
        LDH       *p_to_b,b
        MPY       a,b,prod
        ADD       sum,prod,sum
[count] SUB       count, 1, count
[count] B         loop
        MV        sum,A4
        .endproc  A4, B3                   ; B3 content has been preserved
```

Program 4.9 Linear assembly code

Note: By declaring a register at the beginning and the end of the procedure, and not using it within the procedure, its content will be preserved.

4.4.1 Trip count

From Table 4.10 it is clear that in order to execute the code properly the loop should iterate at least eight times, in other words when the first add occurs. This is known as the minimum trip count. If for instance it is required that the code iterates *n* times, then the loop (code in cycle 8) should be repeated *n* times minus the number of cycles in the epilogue, that is $n - 7$ times. The $n - 7$ is known as the loop count or the trip count.

When the assembly optimiser is invoked, it has no knowledge of the minimum trip count, in order to take the decision as to whether to pipeline the code or not. To help the tools to decide if pipelining is required or not, the programmer can supply the minimum trip count using the directive **`.trip`**. If the minimum trip count is not supplied, the tools will generate both pipelined and un-pipelined code (these are known as redundant loops). At run-time, a branch to the appropriate loop code is performed. The programmer has control to turn off the software pipelining by using the **`-mu`** option when using the compiler, and also has control over avoiding redundant loops by using the **`-ms`** option. When using the **`-ms`** option, a pipelined version of the code is generated if the minimum trip count is smaller than the value specified with the **`.trip`** directive, otherwise the un-pipelined loop code is generated instead. When using the C language, use the **`_n_assert()`** instead of the **`.trip`** directive.

For example, by assembling the programs `FIR1.sa` and `FIR2.sa` and using the **`-k`** option (see Program 4.10) to keep the assembly files (`*.asm`), we can see that the `FIR2.sa` that contains the **`.trip`** directive generates only the pipelined

```
; FIR 1.sa
; This file is assembled by the following
; command:   cl6x -gks -als FIR1.sa -z
.text

FIR        .proc     a0, b0, a1, b2
           .reg      p_to_a, p_to_b, count
           .reg      a, b, prod,sum
           mv        a0,p_to_a
           mv        b0,p_to_b
           mv        a1,count
           zero      sum

loop       ; WITHOUT .trip directive
           ldh       *p_to_a++,a
           ldh       *p_to_b++,b
           mpy       a, b, prod
           add       prod, sum, sum
[count]    sub       count,1,count
[count]    b         loop

           mv        sum,b2
           .endproc  b2
```

(a)

```
; FIR 2.sa
; This file is assembled by the following
; command:   cl6x -gks -als FIR2.sa -z
.text

FIR        .proc     a0, b0, a1, b2
           .reg      p_to_a, p_to_b, to
           .reg      a, b, prod,
           mv        a0,p_to_a
           mv        b0,p_to_b
           mv        a1,count
           zero      sum

loop       .trip 80
           ldh       *p_to_a++, a
           ldh       *p_to_b++, b
           mpy       a, b, prod
           add       prod, sum, sum
[count]    sub       count, 1, count
[count]    b         loop

           mv        sum, b2
           .endproc  b2
```

(b)

Program 4.10 (a) `FIR1.sa` (without `.trip`) and (b) `FIR2.sa` (with `.trip`)

version (see Program 4.11(b)). For the `FIR1.sa` that does not contain the **`.trip`** directive both un-pipelined and pipelined versions are generated (see Program 4.11(a)). The code size has increased when not using the **`.trip`** directive.

```
;***********************************************
;* TMS320C6x ANSI C Codegen          Version 2.00 *
;* Date/Time created:  Thu Feb 11 16:52:12 1999 *
;***********************************************
;***********************************************
;* GLOBAL FILE PARAMETERS                        *
;*                                               *
;*   Architecture          : TMS320C6200         *
;*   Endian                : Little              *
;*   Interrupt Threshold : Disabled              *
;*   Memory Model          : Small               *
;*   Speculative Load      : Disabled            *
;*   Redundant Loops       : Enabled             *
;*   Pipelining            : Enabled             *
;*   Debug Info            : Debug               *
;*                                               *
;***********************************************
FP          .set      A15
DP          .set      B14
SP          .set      B15

            .file     "fir1.sa"

; FIR1.sa
; This file is assembled by the following
; command:    cl6x -gs -als FIR1.sa -z
            .text
            .sect     ".text"
            .align    32
            .sym      FIR,FIR,36,2,0
            .func     7

;***********************************************
;* FUNCTION NAME: FIR                            *
;*                                               *
;* Regs Modified: A0,A3,A4,A5,B0,B1,B2,B4,B5,B6 *
;* Regs Used: A0,A1,A3,A4,A5,B0,B1,B2,B4,B5,B6  *
;***********************************************
FIR:
;**---------------------------------------------*
;
; FIR       .proc     a0, b0, a1, b2
            .sym      p_to_a,21,4,4,32
            .sym      p_to_b,3,4,4,32
            .sym      count,16,4,4,32
;           .reg      p_to_a, p_to_b, count
            .sym      a,20,4,4,32
            .sym      b,0,4,4,32
            .sym      prod,0,4,4,32
            .sym      sum,5,4,4,32
```

(a)

```
;***********************************************
;* TMS320C6x ANSI C Codegen          Version 2.00 *
;* Date/Time created: Thu Feb 11 17:00:59 1999  *
;***********************************************
;***********************************************
;* GLOBAL FILE PARAMETERS                        *
;*                                               *
;*   Architecture          : TMS320C6200         *
;*   Endian                : Little              *
;*   Interrupt Threshold : Disabled              *
;*   Memory Model          : Small               *
;*   Speculative Load      : Disabled            *
;*   Redundant Loops       : Enabled             *
;*   Pipelining            : Enabled             *
;*   Debug Info            : Debug               *
;*                                               *
;***********************************************
FP          .set      A15
DP          .set      B14
SP          .set      B15

            .file     "fir2.sa"

; FIR2.sa
; This file is assembled by the following
; command:    cl6x -gs -als FIR2.sa -z
            .text
            .sect     ".text"
            .align    32
            .sym      FIR,FIR,36,2,0
            .func     7

;***********************************************
;* FUNCTION NAME: FIR                            *
;*                                               *
;* Regs Modified: A0,A3,A4,A5,B0,B2,B4,B5,B6    *
;* Regs Used:  A0,A1,A3,A4,A5,B0,B2,B4,B5,B6    *
;***********************************************
FIR:
;**---------------------------------------------*
;
; FIR       .proc     a0, b0, a1, b2
            .sym      p_to_a,21,4,4,32
            .sym      p_to_b,3,4,4,32
            .sym      count,16,4,4,32
;           .reg      p_to_a, p_to_b, count
            .sym      a,0,4,4,32
            .sym      b,0,4,4,32
            .sym      prod,0,4,4,32
            .sym      sum,5,4,4,32
```

(b)

Program 4.11 Assembly code generated by the tools: (a) from program `FIR1.sa` that does not contain the `.trip` directive; (b) from program `FIR2.sa` that contains the `.trip` directive

Program 4.11 continued

```
;          .reg            a, b, prod,sum
           .line           4
           MV      .L2X    A0,B5        ; |10|
           .line           5
           MV      .L1X    B0,A3        ; |11|
           .line           6
           MV      .L2X    A1,B0        ; |12|
           .line           7
           ZERO    .L1     A5           ; |13|
           CMPGTU  .L2     B0,7,B1
   [ B1]   B       .S1     L4
           NOP             5
           ; BRANCH OCCURS
;**---------------------------------------------*
L2:
           .line           10
           LDH     .D2T2   *B5++,B4     ; |16|
           .line           11
           LDH     .D1T1   *A3++,A0     ; |17|
           NOP             4
           .line           12
           MPY     .M1X    B4,A0,A0     ; |18|
           NOP             1
           .line           13
           ADD     .L1     A0,A5,A5     ; |19|
           .line           14
   [ B0]   SUB     .L2     B0,0x1,B0    ; |20|
           .line           15
   [ B0]   B       .S1     L2           ; |21|
           NOP             5
           ; BRANCH OCCURS              ; |21|
;**---------------------------------------------*
           B       .S1     L9           ; |21|
           NOP             5
           ; BRANCH OCCURS              ; |21|
;**---------------------------------------------*
L4:
           MVC     .S2     CSR,B6       ; |21|
           AND     .L2     -2,B6,B4     ; |21|

           MVC     .S2     B4,CSR       ; |21|
||         SUB     .L2     B0,7,B0      ; |21|
;**---------------------------------------------*
L5:        ; PIPED LOOP PROLOG

           LDH     .D2T2   *B5++,B4     ; |16|
||         LDH     .D1T1   *A3++,A0     ; |17|

   [ B0]   SUB     .L2     B0,0x1,B0    ; |20|
||         LDH     .D2T2   *B5++,B4     ;@ |16|
||         LDH     .D1T1   *A3++,A0     ;@ |17|

   [ B0]   B       .S2     loop         ; |21|
|| [ B0]   SUB     .L2     B0,0x1,B0    ;@ |20|
||         LDH     .D2T2   *B5++,B4     ;@@ |16|
||         LDH     .D1T1   *A3++,A0     ;@@ |17|

   [ B0]   B       .S2     loop         ;@ |21|
|| [ B0]   SUB     .L2     B0,0x1,B0    ;@@ |20|
||         LDH     .D2T2   *B5++,B4     ;@@@ |16|
||         LDH     .D1T1   *A3++,A0     ;@@@ |17|
```

(a)

```
;          .reg            a, b, prod,sum
           .line           4
           MV      .L2X    A0,B5        ; |10|
           .line           5
           MV      .L1X    B0,A3        ; |11|
           .line           6
           MV      .L2X    A1,B0        ; |12|
           .line           7
           ZERO    .L1     A5           ; |13|
           MVC     .S2     CSR,B6
           AND     .L1X    -2,B6,A0

           MVC     .S2X    A0,CSR
||         SUB     .L2     B0,13,B0
;**---------------------------------------------*
L2:        ; PIPED LOOP PROLOG
; loop     .trip  80

           LDH     .D2T2   *B5++,B4     ; |16|
||         LDH     .D1T1   *A3++,A0     ; |17|

           LDH     .D2T2   *B5++,B4     ;@ |16|
||         LDH     .D1T1   *A3++,A0     ;@ |17|

   [ B0]   B       .S2     loop         ; |21|
||         LDH     .D2T2   *B5++,B4     ;@@ |16|
||         LDH     .D1T1   *A3++,A0     ;@@ |17|

   [ B0]   B       .S2     loop         ;@ |21|
||         LDH     .D2T2   *B5++,B4     ;@@@ |16|
||         LDH     .D1T1   *A3++,A0     ;@@@ |17|

   [ B0]   B       .S2     loop         ;@@ |21|
||         LDH     .D2T2   *B5++,B4     ;@@@@ |16|
||         LDH     .D1T1   *A3++,A0     ;@@@@ |17|

           MPY     .M1X    B4,A0,A4     ; |18|
|| [ B0]   B       .S2     loop         ;@@@ |21|
||         LDH     .D2T2   *B5++,B4     ;@@@@@ |16|
||         LDH     .D1T1   *A3++,A0     ;@@@@@ |17|

           MPY     .M1X    B4,A0,A4     ;@ |18|
|| [ B0]   B       .S2     loop         ;@@@@ |21|
||         LDH     .D2T2   *B5++,B4     ;@@@@@@ |16|
||         LDH     .D1T1   *A3++,A0     ;@@@@@@ |17|
;**---------------------------------------------*
loop:      ; PIPED LOOP KERNEL

           ADD     .L1     A4,A5,A5     ; |19|
||         MPY     .M1X    B4,A0,A4     ;@@ |18|
|| [ B0]   B       .S2     loop         ;@@@@@ |21|
|| [ B0]   SUB     .L2     B0,0x1,B0    ;@@@@@@ |20|
||         LDH     .D2T2   *B5++,B4     ;@@@@@@@ |16|
||         LDH     .D1T1   *A3++,A0     ;@@@@@@@ |17|
;**---------------------------------------------*
L4:        ; PIPED LOOP EPILOG

           ADD     .L1     A4,A5,A5     ;@ |19|
||         MPY     .M1X    B4,A0,A4     ;@@@ |18|

           ADD     .L1     A4,A5,A5     ;@@ |19|
||         MPY     .M1X    B4,A0,A4     ;@@@@ |18|
```

(b)

Program 4.11 continued

```
   [ B0] B      .S2     loop        ;@@ |21|
|| [ B0] SUB    .L2     B0,0x1,B0   ;@@@ |20|
||       LDH    .D2T2   *B5++,B4    ;@@@@ |16|
||       LDH    .D1T1   *A3++,A0    ;@@@@ |17|
         MPY    .M1X    B4,A0,A4    ; |18|
|| [ B0] B      .S2     loop        ;@@@ |21|
|| [ B0] SUB    .L2     B0,0x1,B0   ;@@@@ |20|
||       LDH    .D2T2   *B5++,B4    ;@@@@@ |16|
||       LDH    .D1T1   *A3++,A0    ;@@@@@ |17|

         MPY    .M1X    B4,A0,A4    ;@ |18|
|| [ B0] B      .S2     loop        ;@@@@ |21|
|| [ B0] SUB    .L2     B0,0x1,B0   ;@@@@@ |20|
||       LDH    .D2T2   *B5++,B4    ;@@@@@@ |16|
||       LDH    .D1T1   *A3++,A0    ;@@@@@@ |17|

;**--------------------------------------------*
loop:    ; PIPED LOOP KERNEL

         ADD    .L1     A4,A5,A5    ; |19|
||       MPY    .M1X    B4,A0,A4    ;@@ |18|
|| [ B0] B      .S2     loop        ;@@@@@ |21|
|| [ B0] SUB    .L2     B0,0x1,B0   ;@@@@@@ |20|
||       LDH    .D2T2   *B5++,B4    ;@@@@@@@ |16|
||       LDH    .D1T1   *A3++,A0    ;@@@@@@@ |17|

;**--------------------------------------------*
L7:      ; PIPED LOOP EPILOG

         ADD    .L1     A4,A5,A5    ;@ |19|
||       MPY    .M1X    B4,A0,A4    ;@@@ |18|

         ADD    .L1     A4,A5,A5    ;@@ |19|
||       MPY    .M1X    B4,A0,A4    ;@@@@ |18|

         ADD    .L1     A4,A5,A5    ;@@@ |19|
||       MPY    .M1X    B4,A0,A4    ;@@@@@ |18|

         ADD    .L1     A4,A5,A5    ;@@@@ |19|
||       MPY    .M1X    B4,A0,A4    ;@@@@@@ |18|

         ADD    .L1     A4,A5,A5    ;@@@@@ |19|
||       MPY    .M1X    B4,A0,A4    ;@@@@@@@ |18|

         ADD    .L1     A4,A5,A5    ;@@@@@@ |19|
         ADD    .L1     A4,A5,A5    ;@@@@@@@ |19|
;**--------------------------------------------*
         MVC    .S2     B6,CSR      ; |21|
;**--------------------------------------------*
L9:
         .line          17
         MV     .L2X    A5,B2       ; |23|
;**--------------------------------------------*
         .line          18
         .endfunc       24,000000000h,0

;        .endproc       b2
```

(a)

```
         ADD    .L1     A4,A5,A5    ;@@@ |19|
||       MPY    .M1X    B4,A0,A4    ;@@@@@ |18|

         ADD    .L1     A4,A5,A5    ;@@@@ |19|
||       MPY    .M1X    B4,A0,A4    ;@@@@@@ |18|

         ADD    .L1     A4,A5,A5    ;@@@@@ |19|
||       MPY    .M1X    B4,A0,A4    ;@@@@@@@ |18|

         ADD    .L1     A4,A5,A5    ;@@@@@@ |19|
         ADD    .L1     A4,A5,A5    ;@@@@@@@ |19|
;**--------------------------------------------*
         MVC    .S2     B6,CSR
         .line          17
         MV     .L2X    A5,B2       ; |23|
;**--------------------------------------------*
         .line          18
         .endfunc       24,000000000h,0

;        .endproc       b2
```

(b)

Chapter 5

Finite Impulse Response (FIR) filter implementation

5.1 Introduction

Amongst all the obvious advantages that digital filters offer, the Finite Impulse Response (FIR) filter can also guarantee linear phase characteristics that neither analogue nor Infinite Impulse Response (IIR) filters can achieve. There are many commercially available software packages for filter design. However, without basic theoretical knowledge of the FIR filter, it will be difficult to use them.

The main purpose of this chapter is twofold. First, it is to show how to design an FIR filter and implement it on the TMS320C62xx processor, and secondly, it is to show how to optimise code using knowledge gained in Chapter 4. This chapter also shows how to interface C and assembly, how to use intrinsics, and how to put into practice the material that has been covered in the previous chapters.

5.2 Properties of an FIR filter

5.2.1 Filter coefficients

An FIR filter can be defined by the following expression:

$$y[n] = \sum_{k=0}^{N} b_k \cdot x[n-k] \qquad [5.1]$$

where $x[n]$ represents filter input, b_k represents the filter coefficients, and $y[n]$ represents the filter output.

From Equation [5.1], it is clear that any FIR filter can be fully characterised if its coefficients are known. However, there is also a hidden parameter, the sampling rate of the input signal. In the next few sections the sampling rate will be taken into account. So let us look more closely at the coefficients and see what they actually represent. If the signal $x[n]$ is replaced by an impulse $\delta[n]$, Equation [5.1] will lead to

$$\begin{aligned} y[n] &= \sum_{k=0}^{N} b_k \delta[n-k] \\ &= h[n] \end{aligned} \qquad [5.2]$$

where $h[n]$ is the impulse response of the filter. Since

$$\delta[n-k] = \begin{cases} 0 & \text{for } n \neq k \\ 1 & \text{for } n = k \end{cases}$$

Equation [5.2] gives:

$$\begin{aligned} b_0 &= h[0] \\ b_1 &= h[1] \\ &\vdots \\ b_k &= h[k] \end{aligned} \qquad [5.3]$$

We can say that the coefficients of a filter are the same as the impulse response samples of the filter.

5.2.2 Frequency response of an FIR filter

By taking the z-transform of $h[n](H(z))$:

$$H(z) = \sum_{n=0}^{\infty} h[n] z^{-n} \qquad [5.4]$$

To find the frequency response, the parameter z is replaced by $e^{-j\omega}$, therefore:

$$H(z)|_{z=e^{-j\omega}} = H(\omega) = \sum_{n=0}^{\infty} h[n] e^{-jn\omega} \qquad [5.5]$$

Since $e^{-j2\pi n} = 1$, then:

$$H(\omega + 2\pi) = \sum_{n=0}^{\infty} h[n] e^{-jn(\omega+2\pi)} = \sum_{n=0}^{\infty} h[n] e^{-jn\omega}$$

$$H(\omega + 2\pi) = \sum_{n=0}^{\infty} h[n] e^{-jn\omega} = H(\omega)$$

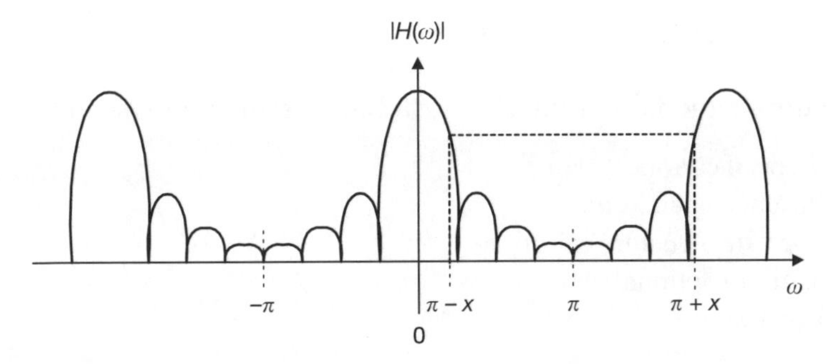

Figure 5.1 An arbitrary frequency response of an FIR filter showing the periodicity

It has therefore been shown that the response of an FIR filter is periodic and its period is 2π (see Figure 5.1).

This property has many practical consequences, one of which is the aliasing effect. For instance, if we choose a signal of frequency $\pi + x$ as shown in Figure 5.1, it will appear at the output of the filter as a signal of frequency $\pi - x$. To avoid aliasing, an analogue filter can be used to remove frequencies beyond the region of interest. Again in a practical situation, this can be very challenging since analogue filters introduce non-linear phase distortion.

5.2.3 Phase linearity of an FIR filter

One of the most important characteristics of an FIR filter is its phase response. The phase response of an FIR filter can be made linear; this is quite important in communications technology.

It can be shown (Parks and Burrus, 1987; Mitra, 1998) that for a causal FIR filter whose impulse response is symmetrical (i.e. $h[n] = h[N-1-n]$ for $n = 0, 1, \dots, N-1$) its phase is guaranteed to be linear. This is summarised in Table 5.1.

Condition	Phase $\left(k = -\frac{N-1}{2}\right)$	Phase property
$h[n] = h[N-n-1]$	$k\omega$	Linear phase
$h[n] = -h[N-n-1]$	$\frac{\pi}{2} + k\omega$	Linear phase and 90° phase shift

Table 5.1 Conditions for linear phase

5.3 Design procedure

To fully design and implement a filter, five main steps are required:

(1) Filter specification
(2) Coefficients calculation
(3) Appropriate structure selection
(4) Simulation (optional)
(5) Implementation.

5.3.1 Specifications

A system can be fully specified by its transfer function, that is magnitude $|H(\omega)|$ and phase $\theta(\omega)$. In general $\theta(\omega)$ can be simply specified to be either

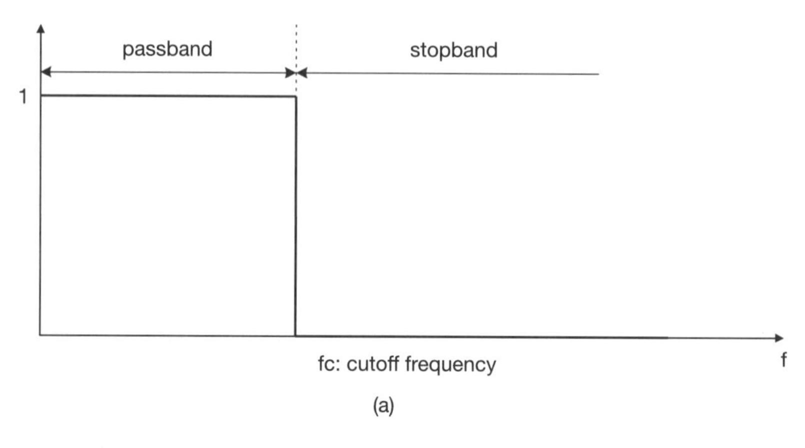

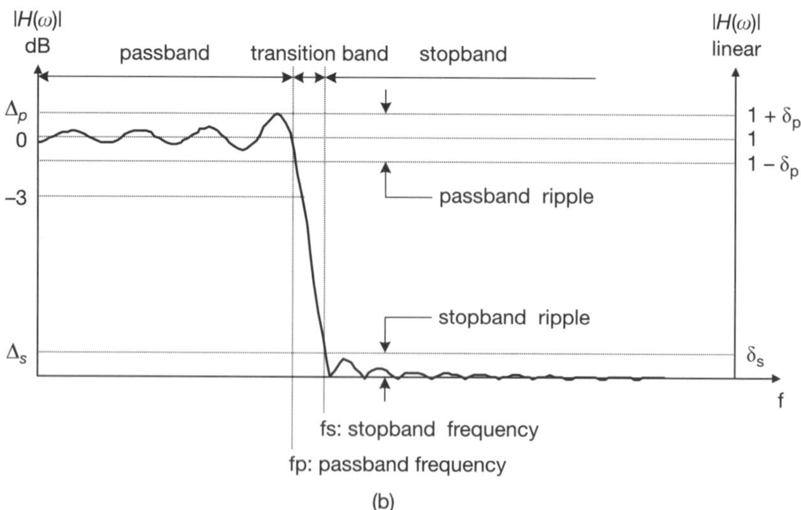

Figure 5.2 Filter specifications: (a) ideal; (b) practical

linear or non-linear. However, the magnitude response is specified by various parameters as shown in Figure 5.2.

The relation between the linear and logarithmic scale is shown below.

$$\Delta p = +20\log_{10}(1 + \delta p)$$

or

$$\delta p = 10^{\Delta p/20} - 1$$
$$\Delta s = +20\log_{10}(\delta s)$$

or

$$\delta s = 10^{\Delta s/20}$$

5.3.2 Coefficients calculation

The ability to achieve a frequency response 'identical' to or within the limits specified will depend mainly on the method used to calculate the coefficients. The most common methods are the window, frequency sampling and Park–McClellan (also known as optimal, equiripple or optimal equiripple). For more details on the design methods the reader is referred to the following references Ludeman (1987), Strum and Kirk (1988), Proakis and Manolakis (1992), Ifeachor and Jervis (1993), McClellan *et al.* (1998) and Mitra (1998). In this chapter the window method has been selected.

5.3.2.1 Window method

To extract the coefficients with this method, four steps are required, each of which is described below.

Step 1: Specify the frequency response (see Figure 5.3)

Step 2: Calculate the coefficients of the impulse response of the ideal filter (Figure 5.4)

$$\begin{aligned} h_d &= \frac{1}{2\pi}\int_{-\pi}^{\pi} H_d(\omega)e^{+j\omega n}d\omega \\ &= \frac{1}{2\pi}\int_{-\omega_c}^{\omega_c} 1 \cdot e^{j\omega n}d\omega \\ &= \begin{cases} \dfrac{2f_c\sin(n\omega_c)}{n\omega_c} & \text{for } n \neq 0 \\ 2f_c & \text{for } n = 0 \end{cases} \end{aligned}$$

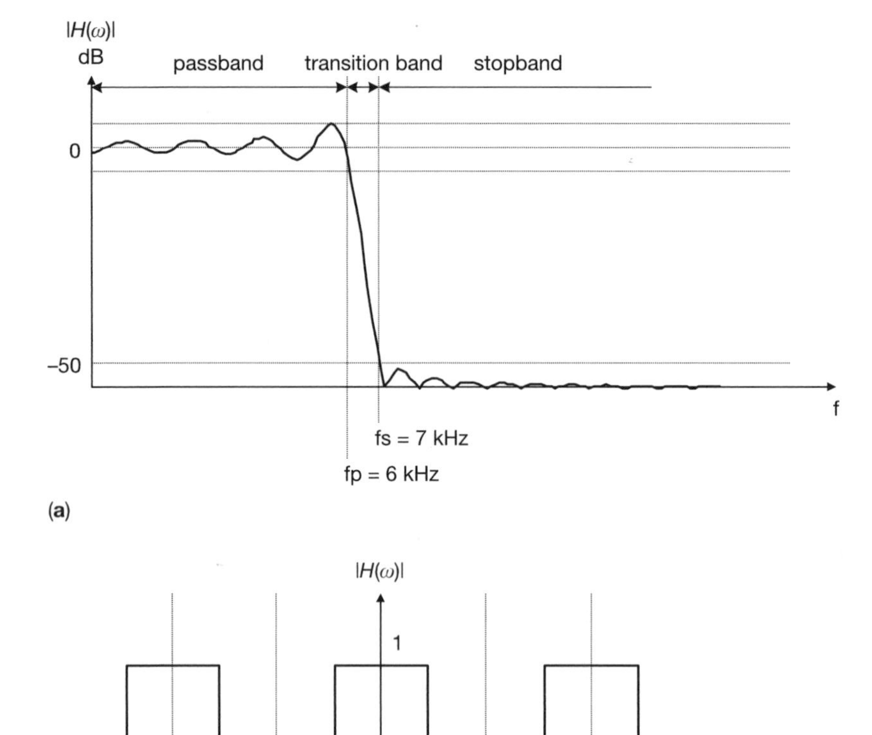

Figure 5.3 Frequency response: (a) desired; (b) ideal

As $\sin(\theta)/\theta = \sin(-\theta)/(-\theta)$, $h_d(n)$ is therefore symmetrical and as a result the phase of the filter is linear. To achieve a magnitude of a transfer function identical to the desired one, the length of the filter should be infinite, which means that in order to implement this filter we need a large amount of time to calculate each output.

To get around this problem the impulse response of the filter should be truncated. Consequently, the effect of truncation introduces overshoots and ripples. This is known as the Gibbs phenomenon. In order to reduce the overshoot and ripple, many window functions have been investigated. Figure 5.5 shows the frequently used windows and Table 5.2 shows the features of each window.

Step 3: Determine the filter length, *N*
For a Hamming window, we have

$$N = \frac{3.3}{\Delta f}$$

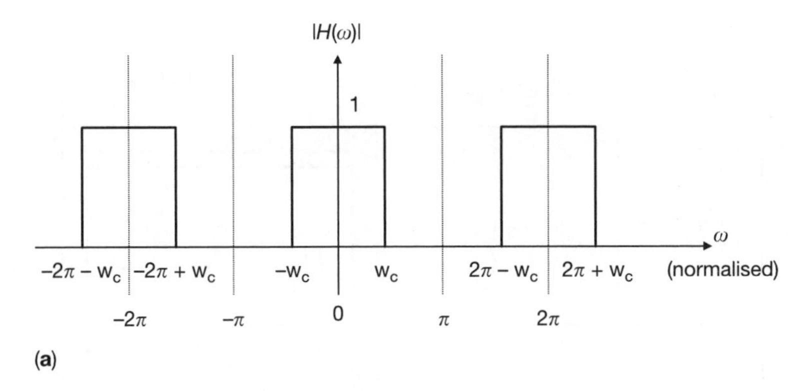

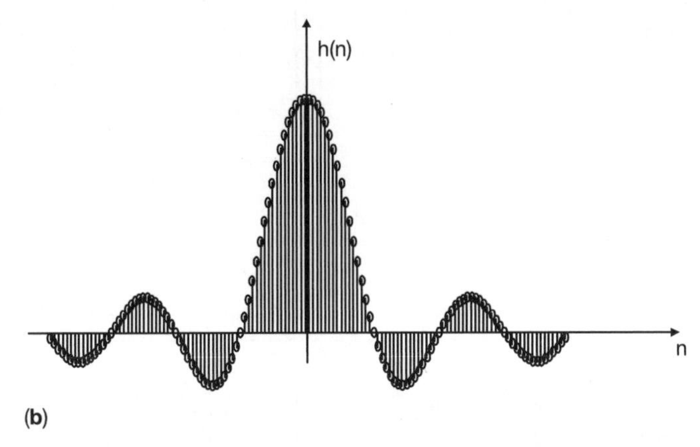

Figure 5.4 (a) Ideal frequency response of a low pass filter and its (b) impulse response

Window type	Normalised transition width (Δf (Hz))	Passband ripple (dB)	Stopband attenuation (dB)
Rectangular	$0.9/N$	0.7416	21
Hanning	$3.1/N$	0.0546	44
Hamming	$3.3/N$	0.0194	53
Blackman	$5.5/N$	0.0017	74

Table 5.2 Window features

where Δf is the normalised transition width; that is $\delta f/f_s$, where f_s is the sampling frequency and δf ($\delta f = f_s - f_p$) is the transition width (see Figure 5.2). Therefore,

$$N = \frac{3.3}{\delta f} f_s = \frac{3.3}{1 \text{ kHz}} \times 40 \text{ kHz} = 132$$

Window's Name	Window Function $w(n), -\frac{(N-1)}{2} \leq n \leq \frac{(N-1)}{2}$	Window Sequence
Rectangular	1	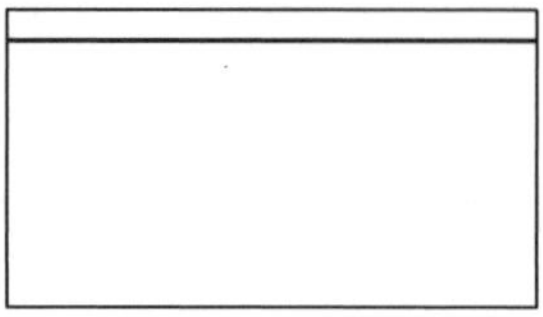
Hanning	$0.5\left[1+\cos\left(\frac{2\pi n}{N}\right)\right]$	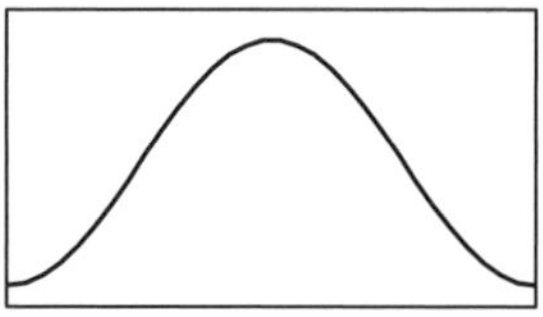
Hamming	$0.54+0.46\cos\left(\frac{2\pi n}{N}\right)$	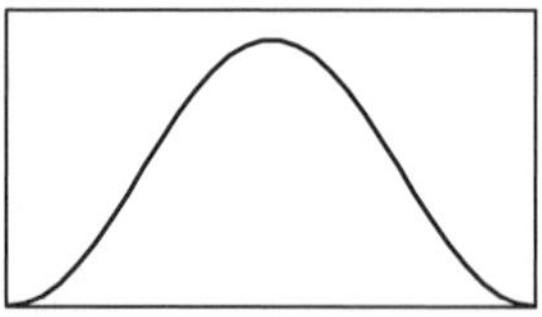
Blackman	$0.42+0.5\cos\left(\frac{2\pi n}{N-1}\right)$ $+0.08\cos\left(\frac{4\pi n}{N-1}\right)$	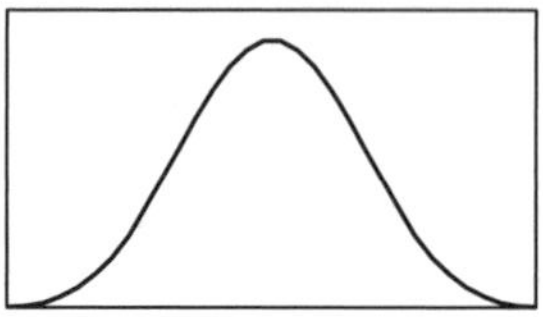

Figure 5.5 Frequently used windows

It is worth observing that the larger the transition width is, the smaller the length of the filter. Let us choose N equal to 133 in order to have odd symmetry for the purpose stated in Section 5.3.3.2.

Step 4: Calculate the set of truncated impulse response coefficients, $\{h(n)\}$

$$h(n) = h_d(n) \cdot W(n) \quad \text{for} \ -\frac{N}{2} \leqslant n \leqslant \frac{N}{2}$$

or

$$W(n) = 0.54 + 0.46\cos\left(\frac{2\pi n}{N}\right) \quad \text{for} \ -\frac{N}{2} \leqslant n \leqslant \frac{N}{2}$$

Since N has been chosen to be 133,

$$W(n) = 0.54 + 0.46\cos\left(\frac{2\pi n}{133}\right) \quad \text{for} \ -66 \leqslant n \leqslant 66$$

The $h(n)$ coefficients have been calculated using a Matlab program (see Program 5.1 and Table 5.3). Program 5.1 also generates the impulse response $h(n)$, plots the frequency response of the filter and saves the coefficients in Q15 format to a file (see Table 5.3, Figure 5.6 and Figure 5.7).

```
fc= 6000/40000;
N=127
%N=133 -1
n=-(N/2):(N/2);
n=n+(n==0)*eps  ; %avoiding division by zero
[h]=sin(n*2*pi*fc)./(n*pi);
[w]= .54 +0.46*cos(2*pi*n/N);
d=h.*w;
d2 =d.*2^15
for i=1:8:N,
fprintf('c:\evm6x\chapter6\fir_coef.txt',' %8.0f, %8.0f, %8.0f, %8.0f, %8.0f, %8.0f,
        %8.0f ,%8.0f\n',d2(i:i+7));
end;

figure(1)
[g,f]=freqz(d,1,512,40000);
plot(f,20*log10(abs(g)));
[g,f]=freqz(d,1,512,40000);
axis([0  2*10^4 -100 1]);
figure(2);
plot(n,d);
figure(3)
axis([0  2 -100 0]);
freqz(d,1,512,40000);
end
```

Program 5.1 MATLAB program for generating the impulse response coefficients

−2	10	14	7	−7	−17	−13	3
19	21	4	−21	−32	−16	18	43
34	−8	−51	−56	−11	53	81	41
−44	−104	−81	19	119	129	24	−119
−178	−88	95	222	171	−41	−248	−266
−50	244	366	181	−195	−457	−353	85
522	568	109	−540	−831	−424	474	1163
953	−245	−1661	−2042	−463	2940	6859	9469
9469	6859	2940	−463	−2042	−1661	−245	953
1163	474	−424	−831	−540	109	568	522
85	−353	−457	−195	181	366	244	−50
−266	−248	−41	171	222	95	−88	−178
−119	24	129	119	19	−81	−104	−44
41	81	53	−11	−56	−51	−8	34
43	18	−16	−32	−21	4	21	19
3	−13	−17	−7	7	14	10	−2

Table 5.3 FIR coefficients converted into Q15 format

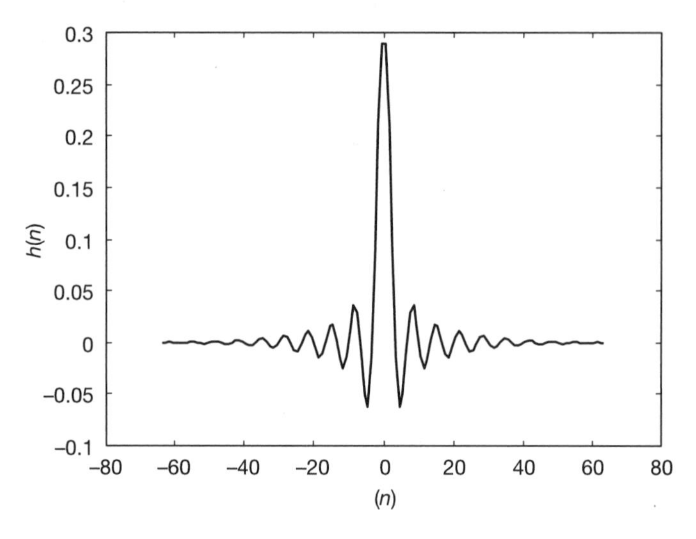

Figure 5.6 Plot of the filter coefficients $h(n)$

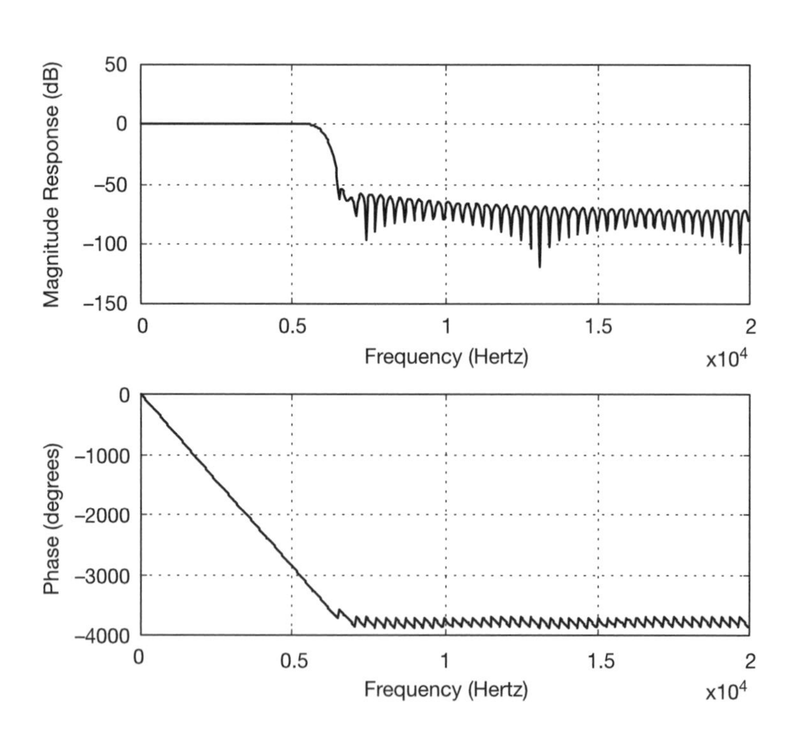

Figure 5.7 Transfer function of the designed filter

5.3.3 Realisation structure

So far, we have been concentrating on the mathematical side in order to calculate the filter coefficients and we found that different methods do exist. To either realise, implement or program a digital filter, different structures exist also. Which structure to use depends on factors such as:

(1) DSP architecture on which the filter is to be implemented;
(2) sensitivity to error in the filter coefficients;
(3) sensitivity to error in the signal.

Structure can be derived by manipulation of the transfer function as shown below. For more details on the structures and factors influencing the choice of one structure over another, the reader is referred to Ludeman (1987) and Mitra (1998).

5.3.3.1 Direct structure

We have seen previously that the transfer function of an FIR filter can be written as follows:

$$H(z) = \sum_{k=0}^{N-1} b_k z^{-k} \tag{5.6}$$

The transfer function can also be expressed as

$$Y(z) = H(z) \cdot X(z) \tag{5.7}$$

where $X(z)$ and $Y(z)$ represent the z-transform of the input and the output sequences.

From Equations [5.6] and [5.7] the difference equation can be derived as

$$y(n) = b_0 x(n) + b_1 x(n-1) + \ldots + b_{N-1} x(n-N+1) \tag{5.8}$$

A 'direct' implementation of Equation [5.8] is illustrated in Figure 5.8. This structure is known as the 'Direct Form' or 'Transversal'.

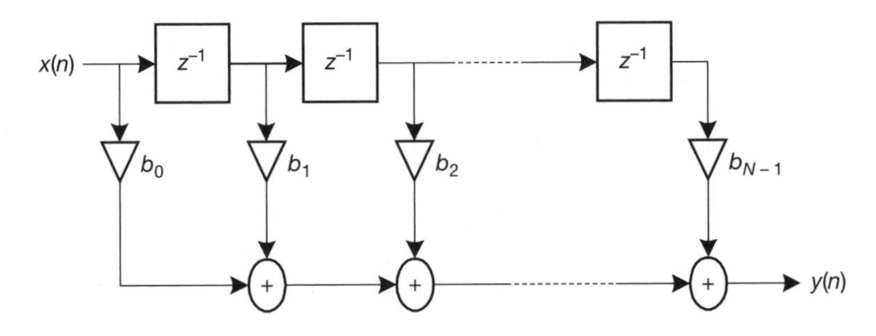

Figure 5.8 Direct form structure for an FIR filter

5.3.3.2 Linear phase structures

By examining Figure 5.8, we can see that in order to implement an FIR filter of length N ($N-1$ taps), N multiplications and $N-1$ additions are required. We have seen previously that in order to have a linear phase the coefficients must be symmetrical, that is $b_0 = b_{N-1}$, $b_1 = b_{N-2}$, etc. Since we are grouping the coefficients in pairs we need to decide if the length of the filter is an even or odd number (see Figure 5.9).

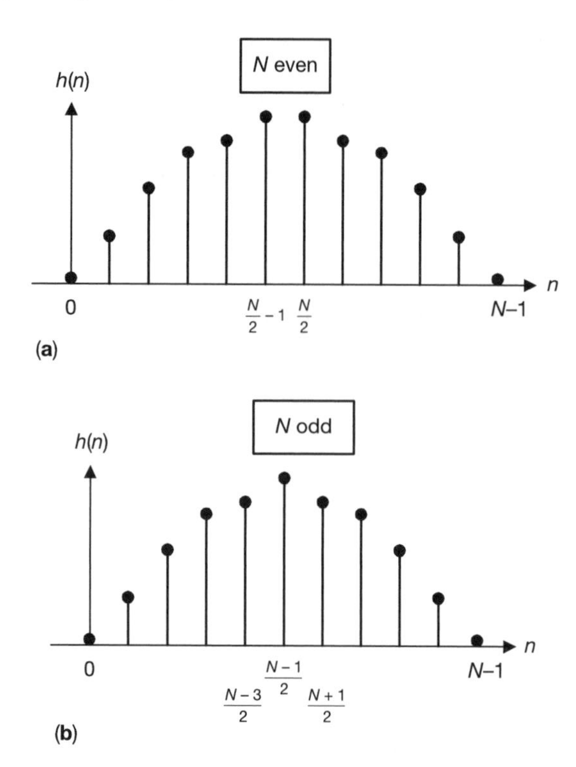

Figure 5.9 Impulse response, *h*(*n*), for: (a) *N* even; (b) *N* odd

If N is even (Figure 5.9(a)) then

$$H(z) = \sum_{k=0}^{N-1} b_k z^{-k} = \sum_{k=0}^{N/2-1} b_k z^{-k} + \sum_{k=N/2}^{N-1} b_k z^{-k}$$

Since the coefficients are symmetrical, then:

$$\begin{aligned} b_0 &= b_{N-1} \\ b_1 &= b_{N-2} \\ b_2 &= b_{N-3} \\ &\vdots \end{aligned}$$

$$b_k = b_{N-k-1}$$
$$\vdots$$
$$b_{\frac{N}{2}-1} = b_{N-(\frac{N}{2}-1)-1} = b_{\frac{N}{2}}$$

Therefore:

$$H(z) = \sum_{k=0}^{\frac{N}{2}-1} b_k (z^{-k} + z^{N-k-1})$$

This leads to the structure shown in Figure 5.10(a). This structure is known as the linear phase structure and it reduces the number of multiplications by half. However, this structure introduces programming complexity since it requires two pointers moving in opposite directions.

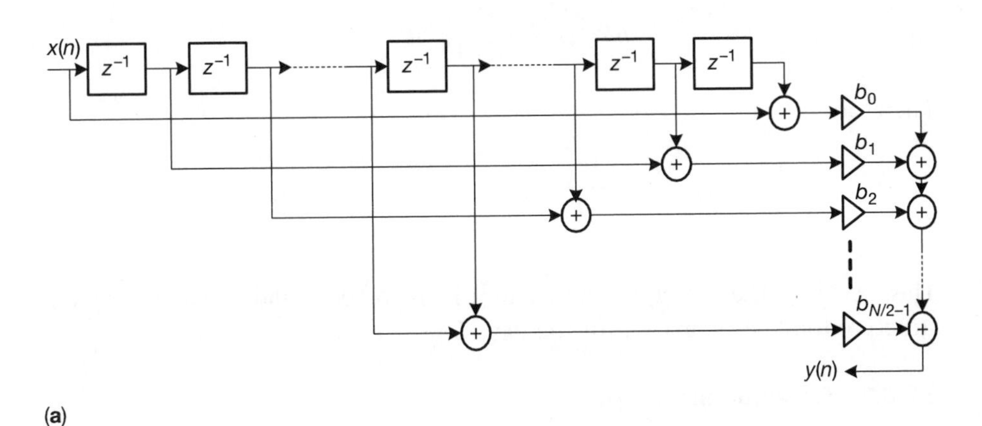

(a)

(b)

Figure 5.10 Linear phase structure for: (a) *N* even; (b) *N* odd

If N is odd then by referring to Figure 5.9(b) for N odd we can write $H(z)$ as follows:

$$H(z) = \sum_{k=0}^{N-1} b_k z^{-k} = \sum_{k=0}^{\frac{N-3}{2}} b_k z^{-k} + b_{\frac{N-1}{2}} z^{\frac{N-1}{2}} + \sum_{k=\frac{N+1}{2}}^{N-1} b_k z^{-k}$$

Since the coefficients are symmetrical, then:

$$\begin{aligned} b_0 &= b_{N-1} \\ b_1 &= b_{N-2} \\ b_2 &= b_{N-3} \\ &\vdots \\ b_k &= b_{N-k-1} \\ &\vdots \\ b_{\frac{N-3}{2}} &= b_{N-\left(\frac{N-3}{2}\right)-1} = b_{\frac{N+1}{2}} \end{aligned}$$

Therefore

$$H(z) = \sum_{k=0}^{\frac{N-3}{2}} b_k (z^{-k} + z^{N-k-1}) + b_{\frac{N-1}{2}} z^{\frac{N-1}{2}}$$

This leads to the structure shown in Figure 5.10(b) and is similar to the structure for N even except for the term $b_{\frac{N-1}{2}}$.

5.3.3.3 Cascade structures

The cascade structure converts the transfer function into a product of second-order functions, as follows:

$$\begin{aligned} H(z) &= \sum_{k=0}^{N-1} b_k z^{-k} = b_0 + b_1 z^{-1} + b_2 z^{-2} + \ldots + b_{N-1} z^{-(N-1)} \\ &= b_0 \left[1 + \frac{b_1}{b_0} z^{-1} + \frac{b_2}{b_0} z^{-2} + \ldots + \frac{b_{N-1}}{b_0} z^{-(N-1)} \right] \end{aligned}$$

We know that the zeros occur in complex conjugate pairs in order to have a linear phase, therefore $H(z)$ can be rewritten as follows:

$$H(z) = b_0 \prod_{k=1}^{\frac{N}{2}} (1 + b_{k,1} z^{-1} + b_{k,2} z^{-2})$$

This leads to the cascade form shown in Figure 5.11.

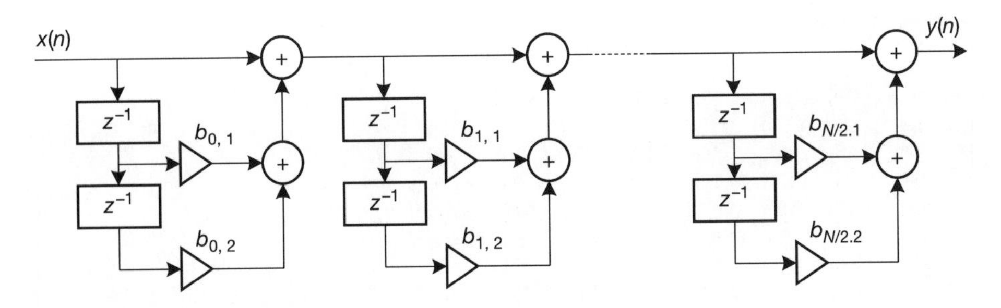

Figure 5.11 Cascade structure

5.3.4 Filter implementation

Once the filter coefficients have been calculated and a structure selected (step 3), the filter can be implemented (step 5). To make the implementation simple let us first implement the filter designed in Section 5.3 using the direct form structure. The difference equation of the filter has been shown to be

$$y(n) = b_0x_0 + b_1x_1 + \ldots + b_{N-1}x_{N-1}$$

This equation can be converted into C code as shown in Program 5.2.

```
/*---------------------------------------------------------------------
    File:          fir.c
    Description: Perform 128-tap fir filtering on codec input.
    Input:         Right channel of line-in
    Output:        Right channel = original signal
                   Left channel  = filtered signal
---------------------------------------------------------------------*/

#include <stdio.h>
#include <string.h>
#include <common.h>
#include <codec.h>
#include <mcbspdrv.h>
#include <board.h>
#include <stdlib.h>
#include <timer.h>
#include <intr.h>
#include <regs.h>

#define SAMPLING_RATE 44100

extern int inicodec (int sampling_rate);

short R_in[128];          // Input samples R_in[0] most recent, R_in[127] oldest.
short h[]=                // Impulse response of FIR filter.
```

Program 5.2 Complete C program for an FIR filter

Program 5.2 continued

```
{
    -2,    10,    14,     7,    -7,   -17,   -13,     3,
    19,    21,     4,   -21,   -32,   -16,    18,    43,
    34,    -8,   -51,   -56,   -11,    53,    81,    41,
   -44,  -104,   -81,    19,   119,   129,    24,  -119,
   178,   -88,    95,   222,   171,   -41,  -248,  -266,
   -50,   244,   366,   181,  -195,  -457,  -353,    85,
   522,   568,   109,  -540,  -831,  -424,   474,  1163,
   953,  -245, -1661, -2042,  -463,  2940,  6859,  9469,
  9469,  6859,  2940,  -463, -2042, -1661,  -245,   953,
  1163,   474,  -424,  -831,  -540,   109,   568,   522,
    85,  -353,  -457,  -195,   181,   366,   244,   -50,
  -266,  -248,   -41,   171,   222,    95,   -88,  -178,
  -119,    24,   129,   119,    19,   -81,  -104,   -44,
    41,    81,    53,   -11,   -56,   -51,    -8,    34,
    43,    18,   -16,   -32,   -21,     4,    21,    19,
     3,   -13,   -17,    -7,     7,    14,    10,    -2
};

#pragma CODE_SECTION (fir_filter, ".iprog")  // Put routine in internal memory.
short fir_filter (short input)
{
  int i;
  short output;
  int acc=0;
  int prod;

  R_in[0] = input;                           // Update most recent sample.

  acc = 0;                                   // Zero accumulator.
  for (i=0; i<128; i++)                      // 128 taps
  {
    prod = (h[i]*R_in[i]);                   // perform Q.15 multiplication
    acc = acc + prod;                        // Update 32-bit accumulator, catering
  }                                          // for temporary overflow.
  output = (short) (acc>>15);                // Cast output to 16-bits.

  for (i=127; i>0; i--)                      // Shift delay samples.
    R_in[i]=R_in[i-1];

  return output;
}

#pragma CODE_SECTION (isr_rint0, ".iprog")   // Put routine in internal memory.
interrupt void isr_rint0 (void)
{
  int sample;
  short output;
  sample = MCBSP0_DRR;
  sample &= 0xffff;

  output = fir_filter ((short) sample);

  sample |= (((int)output)<<16);
  MCBSP0_DXR = sample;
  return;
}
```

Program 5.2 continued

```
void Init_Interrupts ()
{
  intr_reset();                      // Reset the interrupt system,
                                     // disable all interrupts.
  INTR_CLR_FLAG (CPU_INT15);         // Clear previous interrupt request
  INTR_ENABLE (CPU_INT15);           // Enable cpu interrupt line 15
  intr_map(CPU_INT15,ISN_RINT0);     // Link receive interrupt of serial
                                     // port 0 to interrupt line 15.
  intr_hook (isr_rint0,CPU_INT15);   // Assign the interrupt service routine
}

void main(void)
{
  evm_init();                        // Initialise EVM board.
  inicodec (SAMPLING_RATE);          // Initialise codec, adjust sampling rate.
  Init_Interrupts ();                // Initialise interrupts and hook isr.
  INTR_GLOBAL_ENABLE();              // Enable global interrupt.

  for (;;);                          // Main loop, does nothing.
}
```

The set-up used to test the filter is shown in Figure 5.12(a). The transfer function of the designed filter is shown in Figure 5.12(b) (the program should be compiled using `-o3` option).

By using the profiler, or simply by using two breakpoints and the '`runb`' command to count the number of cycles between breakpoints, the code can be benchmarked. In this example, the filter took 7703 cycles for the computation of one output, $y(n)$ and shifting the input samples. In the next section, the FIR

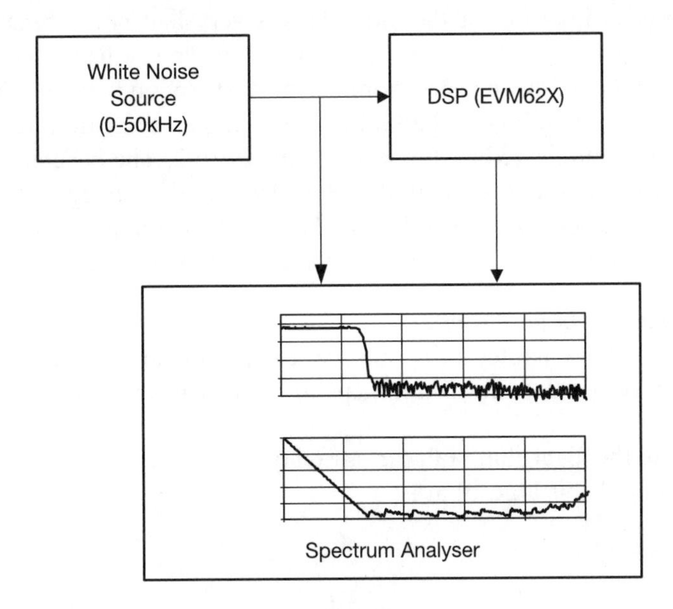

Figure 5.12(a) Set-up for testing the FIR filter

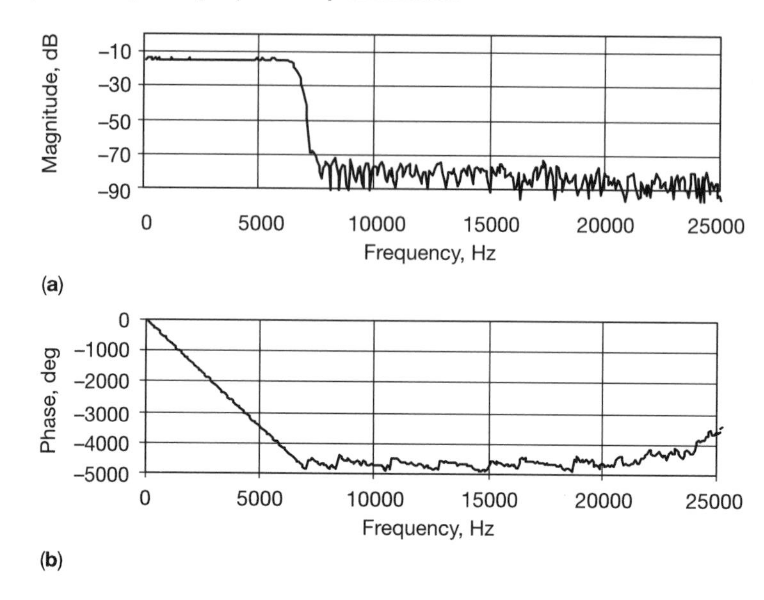

Figure 5.12(b) Frequency response of the designed filter: (a) magnitude; (b) phase

code is optimised using the technique shown in Chapter 4. See directory `F:\DSPCODE\Fir (Chap5)\C\` for the complete code.

5.3.4.1 Filter optimisation

By benchmarking the main part of the FIR code, we find that it takes 32 cycles to perform each iteration of the loop. This is very bad news, because lower performance processors can achieve this in one cycle (by using the multiply, accumulate and data move instruction). In the next section it will be shown that this can be achieved in one or half a cycle. When the `-o3` optimisation level is used, the 7703 figure is reduced to 427 and the 32 to 2. The benchmark can be done by setting a breakpoint at the '`for`' level and then opening a watch window for the clock (`wa clk`) and using the '`runb`' command (see Figure 5.13), or by directly examining the assembly code generated by the compiler.

Hand optimisation of an FIR filter

It was suggested in Chapter 4 that in order to optimise an algorithm, five steps are required. So let us follow these steps to implement the FIR filter.

Step 1: Write the algorithm in linear assembly

The C version of this algorithm is:

```
for (i=0;i<n;i++)
{
  acc = acc + ((_mpy (h[i],R_in[i])) <<1);
}
```

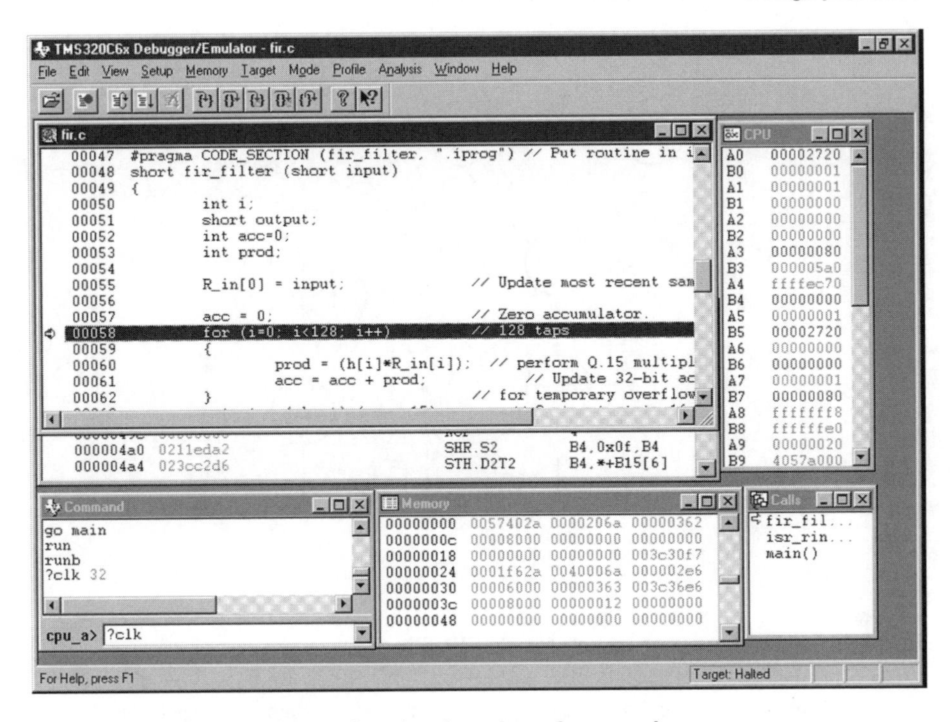

Figure 5.13 Screen dump showing benchmark procedure

This translates to:

```
loop     LDH  *p_to_a,a
         LDH  *p_to_b,b
         MPY  a,b,prod
         SHL  prod,1,prod
         ADD  sum,prod,sum
         SUB  count,1,count
[count]  B    loop
```

Step 2: Create a dependency graph
See Figure 5.14

Step 3: Resource allocation
This is similar to Table 4.8 in Chapter 4 except for .S1. It is shown that the resources have not been exceeded, so it is possible that the kernel of the algorithm can be executed in one cycle. The register allocation can be done at this stage and is shown in Figure 5.15.

Step 4: Scheduling table
The longest path takes nine cycles $(5+2+1+1)$ as shown in Figure 5.16. Therefore the scheduling table is nine cycles long, as shown in Table 5.4.

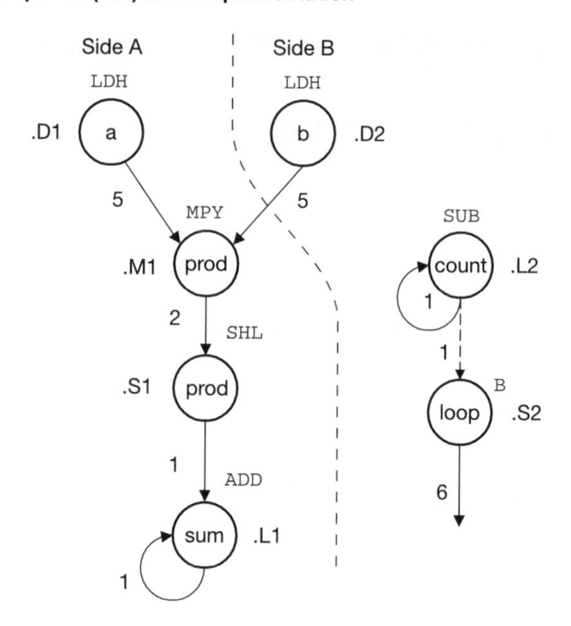

Figure 5.14 Dependency graph for an FIR filter

Register file A	Symbolic registers
A0	
A1	&a
A2	a
A3	prod
A4	sum
A5	

Symbolic registers	Register file B
count	B0
&b	B1
b	B2
	B3
	B4
	B5

Figure 5.15 Register allocation

	Prologue								Loop
Cycle/unit	**1**	**2**	**3**	**4**	**5**	**6**	**7**	**8**	**9**
.D1	LDH	LDH	LDH	LDH	LDH	LDH	LDH	LDH	**LDH**
.D2	LDH	LDH	LDH	LDH	LDH	LDH	LDH	LDH	**LDH**
.L1									**ADD**
.L2			SUB	SUB	SUB	SUB	SUB	SUB	**SUB**
.S1								SHL	**SHL**
.S2				B	B	B	B	B	**B**
.M1						MPY	MPY	MPY	**MPY**
.M2									

Table 5.4 Scheduling table

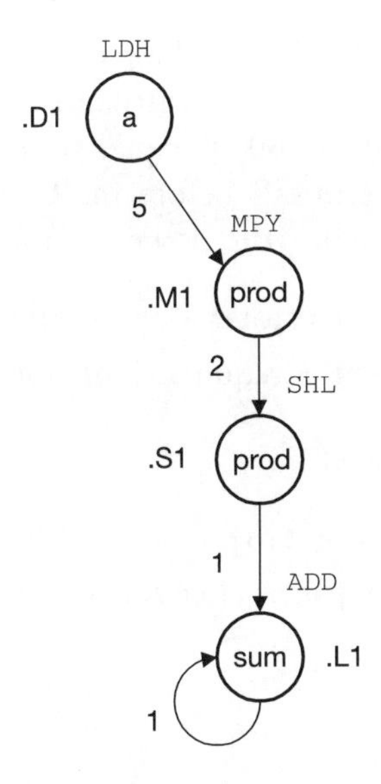

Figure 5.16 Longest path of the dependency graph shown in Figure 5.14

Let us generate a second scheduling table (Table 5.5), which generates the epilogue. This will be useful since when using circular buffers, no extra load will be allowed.

Step 5: Code generation

From Table 5.5 we can now generate the assembly code. This is shown in Program 5.3. It is clear from the scheduling table and Program 5.3 that the kernel of the filter is now executed in one cycle. What is required now is to determine how many times the loop has to iterate. The code has to iterate 128 times, which also means that 128 loads for each unit are needed. Since in the prologue eight loads from each unit are required, the loop has to be repeated 120 times (128 – 8).

	Prologue								Loop	Epilogue							
Cycle/ unit	**1**	**2**	**3**	**4**	**5**	**6**	**7**	**8**	**9**	**10**	**11**	**12**	**13**	**14**	**15**	**16**	**17**
.D1	LDH	LDH	LDH	LDH	LDH	LDH	LDH	LDH	**LDH**								
.D2	LDH	LDH	LDH	LDH	LDH	LDH	LDH	LDH	**LDH**								
.L1									**ADD**	ADD	ADD	ADD	ADD	ADD	ADD	ADD	ADD
.L2			SUB	SUB	SUB	SUB	SUB	SUB	**SUB**								
.S1								SHL	**SHL**	SHL	SHL	SHL	SHL	SHL	SHL	SHL	
.S2				B	B	B	B	B	**B**								
.M1						MPY	MPY	MPY	**MPY**	MPY	MPY	MPY	MPY	MPY			
.M2																	

Table 5.5 Scheduling table with the epilogue

At first glance, we may think that the loop counter should also be set to 120. This can only be true if the branch instruction has no delay states. If we examine the loop kernel when 'B0' is equal to zero, the loop kernel keeps executing for another six iterations before the branch occurs. If we want to find out the loop count, we can write down the following equations:

$$\text{Required loop kernel count} = \text{total loop iterations} - \text{length of prologue}$$
$$\text{Actual loop kernel count} = \text{required loop kernel count} - 6$$

In the above example we have:

$$\text{Required loop kernel count} = 128 - 8 = 120$$
$$\text{Actual loop kernel count} = 120 - 6 = 114$$

```
; PIPED LOOP PROLOGUE
            LDH     .D2T2   *B5++,B4
 ||         LDH     .D1T1   *A7++,A0

            LDH     .D2T2   *B5++,B4
 ||         LDH     .D1T1   *A7++,A0

[ B0]       SUB     .L2     B0,0x1,B0
 ||         LDH     .D2T2   *B5++,B4
 ||         LDH     .D1T1   *A7++,A0

    [ B0]   B       .S2     loop
 || [ B0]   SUB     .L2     B0,0x1,B0
 ||         LDH     .D2T2   *B5++,B4
 ||         LDH     .D1T1   *A7++,A0

    [ B0]   B       .S2     loop
 || [ B0]   SUB     .L2     B0,0x1,B0
 ||         LDH     .D2T2   *B5++,B4
 ||         LDH     .D1T1   *A7++,A0
    [ B0]   B       .S2     loop
 || [ B0]   SUB     .L2     B0,0x1,B0
 ||         LDH     .D2T2   *B5++,B4
 ||         LDH     .D1T1   *A7++,A0

            MPY     .M1X    B4,A0,A4
 || [ B0]   B       .S2     loop
 || [ B0]   SUB     .L2     B0,0x1,B0
 ||         LDH     .D2T2   *B5++,B4
 ||         LDH     .D1T1   *A7++,A0

            MPY     .M1X    B4,A0,A4
 || [ B0]   B       .S2     loop
 || [ B0]   SUB     .L2     B0,0x1,B0
 ||         LDH     .D2T2   *B5++,B4
 ||         LDH     .D1T1   *A7++,A0
 ||         SHL     .S1     A4,1,A4
```

Program 5.3 Hand-optimised assembly code for an FIR filter

Program 5.3 continued

```
;** -----------------------------------------------------------------------*
loop:    ; PIPED LOOP KERNEL

           ADD     .L1     A4,A5,A5
  ||       MPY     .M1X    B4,A0,A4
  || [ B0] B       .S2     loop
  || [ B0] SUB     .L2     B0,0x1,B0
  ||       LDH     .D2T2   *B5++,B4
  ||       LDH     .D1T1   *A7++,A0
  ||       SHL     .S1     A4,1,A4
;** -----------------------------------------------------------------------*
PIPED LOOP EPILOGUE

           ADD     .L1     A4,A5,A5
  ||       MPY     .M1X    B4,A0,A4
  ||       SHL     .S1     A4,1,A4

           ADD     .L1     A4,A5,A5
  ||       MPY     .M1X    B4,A0,A4
  ||       SHL     .S1     A4,1,A4

           ADD     .L1     A4,A5,A5
  ||       MPY     .M1X    B4,A0,A4
  ||       SHL     .S1     A4,1,A4

           ADD     .L1     A4,A5,A5
  ||       MPY     .M1X    B4,A0,A4
  ||       SHL     .S1     A4,1,A4

           ADD     .L1     A4,A5,A5
  ||       MPY     .M1X    B4,A0,A4
  ||       SHL     .S1     A4,1,A4

           ADD     .L1     A4,A5,A5
  ||       SHL     .S1     A4,1,A4

           ADD     .L1     A4,A5,A5
  ||       SHL     .S1     A4,1,A4

           ADD     .L1     A4,A5,A5
```

Linear assembly program

By adding the 'SHL' instruction into Program 4.9 in Chapter 4 we obtain Program 5.4.

```
; FIR.sa
; This file is assembled by the following command:
; cl6x -gs -als FIR.sa -z
        .text

FIR     .proc   a0, b0, a1, b2         ; Do not need to follow the register conventions
        .reg    p_to_a, p_to_b, count  ; if not interfacing to C.
        .reg    a, b, prod,sum
```

Program 5.4 Linear assembly code for an FIR filter

Program 5.4 continued

```
         mv      a0,p_to_a
         mv      b0,p_to_b
         mv      a1,count
         zero    sum

loop     .trip   128
         ldh     *p_to_a++,a
         ldh     *p_to_b++,b
         mpy     a, b, prod
         shl     prod,1,prod
         add     prod, sum, sum
[count]  sub     count,1,count
[count]  b       loop

         mv      sum,b2
        .endproc b2
```

By using the '-mk', we can tell the compiler that no extra loads are permitted so that an epilogue will be generated. The following command line is used to generate the code shown in Program 5.5:

```
Cl6x -gs -mk0  fir.sa
```

Note: No space should be inserted between mk and 0.

Note that the only difference between the code generated by 'hand' and the code generated by the linear assembly is that there is no SUB instruction in the prologue. However, this has been taken into account by setting the loop kernel count to 120 minus 6 (the number 6 comes from the six SUB instructions removed from the prologue).

```
;****************************************************************
;* TMS320C6x ANSI C Codegen                        Version 2.00 *
;* Date/Time created: Fri Mar  5 21:25:01 1999                  *
;****************************************************************

;****************************************************************
;* GLOBAL FILE PARAMETERS                                       *
;*                                                              *
;*   Architecture        : TMS320C6200                          *
;*   Endian              : Little                               *
;*   Interrupt Threshold : Disabled                             *
;*   Memory Model        : Small                                *
;*   Speculative Load    : Disabled                             *
;*   Redundant Loops     : Enabled                              *
;*   Pipelining          : Enabled                              *
;*   Debug Info          : Debug                                *
;*                                                              *
;****************************************************************
```

Program 5.5 Assembly code generated by the tools from the linear assembly

Program 5.5 continued

```
FP      .set    A15
DP      .set    B14
SP      .set    B15

        .file   "fir.sa"

; FIR.sa
; This file is assembled by the following command:
; cl6x -gs -als FIR.sa -z
        .text
        .sect   ".text"
        .align  32
        .sym    FIR,FIR,36,2,0
        .func   7
;****************************************************************
;* FUNCTION NAME: FIR                                           *
;*                                                              *
;*   Regs Modified     : A0,A3,A4,A5,A6,B0,B2,B4,B5,B6          *
;*   Regs Used         : A0,A1,A3,A4,A5,A6,B0,B2,B4,B5,B6       *
;****************************************************************
FIR:
;** -----------------------------------------------------------*
;
; FIR     .proc    a0, b0, a1, b2
          .sym     p_to_a,21,4,4,32
          .sym     p_to_b,4,4,4,32
          .sym     count,16,4,4,32
;         .reg     p_to_a, p_to_b, count
          .sym     a,0,4,4,32
          .sym     b,0,4,4,32
          .sym     prod,0,4,4,32
          .sym     sum,3,4,4,32
;         .reg     a, b, prod,sum
          .line    4
          MV       .L2X    A0,B5        ; |10|
          .line    5
          MV       .L1X    B0,A4        ; |11|
          .line    6
          MV       .L2X    A1,B0        ; |12|
          .line     7
          ZERO     .L1     A3           ; |13|
          MVC      .S2     CSR,B6
          AND      .L1X    -2,B6,A0
          MVC      .S2X    A0,CSR
||        SUB      .L2     B0,14,B0

;** -----------------------------------------------------------*
L2:        ; PIPED LOOP PROLOG
; loop    .trip    128

          LDH      .D2T2   *B5++,B4     ; |16|
||        LDH      .D1T1   *A4++,A0     ; |17|

          LDH      .D2T2   *B5++,B4     ;@ |16|
||        LDH      .D1T1   *A4++,A0     ;@ |17|
```

Program 5.5 continued

```
        LDH     .D2T2   *B5++,B4      ;@@ |16|
||      LDH     .D1T1   *A4++,A0      ;@@ |17|

   [ B0] B      .S2     loop          ; |22|
||      LDH     .D2T2   *B5++,B4      ;@@@ |16|
||      LDH     .D1T1   *A4++,A0      ;@@@ |17|

   [ B0] B      .S2     loop          ;@ |22|
||      LDH     .D2T2   *B5++,B4      ;@@@@ |16|
||      LDH     .D1T1   *A4++,A0      ;@@@@ |17|

        MPY     .M1X    B4,A0,A6      ; |18|
|| [ B0] B      .S2     loop          ;@@ |22|
||      LDH     .D2T2   *B5++,B4      ;@@@@@ |16|
||      LDH     .D1T1   *A4++,A0      ;@@@@@ |17|

        MPY     .M1X    B4,A0,A6      ;@ |18|
|| [ B0] B      .S2     loop          ;@@@ |22|
||      LDH     .D2T2   *B5++,B4      ;@@@@@@ |16|
||      LDH     .D1T1   *A4++,A0      ;@@@@@@ |17|

        SHL     .S1     A6,0x1,A5     ; |19|
||      MPY     .M1X    B4,A0,A6      ;@@ |18|
|| [ B0] B      .S2     loop          ;@@@@ |22|
||      LDH     .D2T2   *B5++,B4      ;@@@@@@@ |16|
||      LDH     .D1T1   *A4++,A0      ;@@@@@@@ |17|

;** ----------------------------------------------------------*
loop:       ; PIPED LOOP KERNEL

        ADD     .L1     A5,A3,A3      ; |20|
||      SHL     .S1     A6,0x1,A5     ;@ |19|
||      MPY     .M1X    B4,A0,A6      ;@@@ |18|
|| [ B0] B      .S2     loop          ;@@@@@ |22|
|| [ B0] SUB    .L2     B0,0x1,B0     ;@@@@@@ |21|
||      LDH     .D2T2   *B5++,B4      ;@@@@@@@@ |16|
||      LDH     .D1T1   *A4++,A0      ;@@@@@@@@ |17|

;** ----------------------------------------------------------*
L4:     ; PIPED LOOP EPILOG

        ADD     .L1     A5,A3,A3      ;@ |20|
||      SHL     .S1     A6,0x1,A5     ;@@ |19|
||      MPY     .M1X    B4,A0,A6      ;@@@@ |18|

        ADD     .L1     A5,A3,A3      ;@@ |20|
||      SHL     .S1     A6,0x1,A5     ;@@@ |19|
||      MPY     .M1X    B4,A0,A6      ;@@@@@ |18|

        ADD     .L1     A5,A3,A3      ;@@@ |20|
||      SHL     .S1     A6,0x1,A5     ;@@@@ |19|
||      MPY     .M1X    B4,A0,A6      ;@@@@@@ |18|

        ADD     .L1     A5,A3,A3      ;@@@@ |20|
||      SHL     .S1     A6,0x1,A5     ;@@@@@ |19|
||      MPY     .M1X    B4,A0,A6      ;@@@@@@@ |18|

        ADD     .L1     A5,A3,A3      ;@@@@@ |20|
```

Program 5.5 continued

```
||      SHL     .S1     A6,0x1,A5     ;@@@@@@ |19|
||      MPY     .M1X    B4,A0,A6      ;@@@@@@@@ |18|

        ADD     .L1     A5,A3,A3      ;@@@@@@ |20|
||      SHL     .S1     A6,0x1,A5     ;@@@@@@@ |19|

        ADD     .L1     A5,A3,A3      ;@@@@@@@ |20|
||      SHL     .S1     A6,0x1,A5     ;@@@@@@@@ |19|

        ADD     .L1     A5,A3,A3      ;@@@@@@@@ |20|
;** ----------------------------------------------------------*
        MVC     .S2     B6,CSR
        .line   18
        MV      .L2X    A3,B2         ; |24|
;** ----------------------------------------------------------*
        .line   19
        .endfunc 25,000000000h,0

;       .endproc b2
```

5.3.4.2 Circular addressing

In order to implement a filter, we also need to shift the data (or update the samples) when the sum of the product function is completed, as shown in the following example.

```
for (i=0;i<n;i++)
  {
    acc = acc + ((_mpy (h[i],R_in[i])) <<1);
  }

  for (j=n-1;j>0;j--)          // update the samples
    {R_in[j]=R_in[j-1];}
```

This process of shifting the data is time consuming and can be avoided by using the circular addressing as shown below.

Circular addressing

Three steps are required in order to set the circular buffers with the 'C6x processors. These steps are described below.

Step 1: Select which pointer(s) need to be used as a circular pointer

There are eight registers that can be used as circular pointers: A4–A7 and B4–B7. To program these registers to operate in a circular mode, select the appropriate mode bits in the Address Mode Register (AMR) as shown in Figure 5.17.

Step 2: Determine the block size of the circular buffer

There are two block sizes, BK0 and BK1, that can be selected. The block size can only be a power of 2, as shown in Equation [5.9]. The parameter N is loaded into the AMR as shown in Figure 5.17.

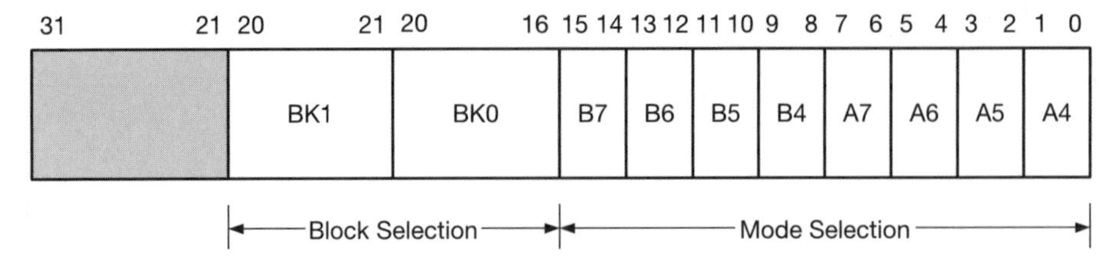

Figure 5.17 Address Mode Register (AMR)

$$\text{Block size (bytes)} = 2^{N+1} \quad [5.9]$$

The maximum value of N is 31 and therefore the maximum block size is 2^{32} bytes. It is important at this stage to note that it is not practical for the block size to be always a power of 2. For instance, if the algorithm has a block size of 150 then the nearest programmable block size is 256, a difference of 106, which means that up to 106 cycles could be lost per iteration.

Step 3: Align the data on the buffer size boundary

When using circular buffers, data has to be aligned on the buffer size boundary. This is easily achieved by using the `#pragma` directive as follows:

```
#pragma DATA_ALIGN (symbol, constant(bytes))
```

Now that the three steps have been defined, let us apply them to the filter designed in Section 5.3.2.

Step 1: Choose B5 as a pointer to the coefficients $h[n]$ and A7 as a pointer to the input sample $x[n]$ (see Figure 5.18).

Step 2: Let us choose BK0 as the holder of the block size. From previous calculations it has been shown that the filter length is 132, but due to the restriction on the buffer sizes shown above, the length of the filter has to be changed to the nearest allowed block size, this being 128.

Steps 1 and 2 can be translated into assembly code by the following assembly code sequence:

```
MVKL    0x0440,B2  ; B5 and A7 used for circular addressing
MVKLH   0x7,B2     ; N=7 -> 2^N+1 = 128 *2
MVC     B2,AMR
```

The block size is defined in bytes, and since the coefficients and the input data are held in 16-bit memories then the buffer size is changed to 256.

Step 3: The arrays $h[n]$ and $x\_in[n]$ have been declared as *short* as follows:

```
short h[]={ -2, 10, 14, 7, -7, ... };

short x_in[128];
```

Each array must be aligned to a 256-byte boundary. This alignment could be achieved by

```
#pragma DATA_ALIGN(h,256)
#pragma DATA_ALIGN(x_in,256)
```

Setting the initial values of the circular pointers

When the first interrupt occurs, the FIR filter algorithm is called, the pointer *B5 should be pointing to $h[0]$ and the *A7 pointer should be pointing to $x[0]$ as shown in Figure 5.18. After 128 loop iterations the output of the filter, acc0, should be:

$$\mathrm{acc0} = h_0 x_0 + h_{n-(N-1)} x_1 + \ldots + h_1 x_{n-(N-1)}$$

When the second interrupt occurs, the newest sample needs to overwrite the oldest sample, that is x_1, along with the *A7 pointer pointing to it, whereas *B5 should point to $h[0]$. There is no problem with *B5 pointing to $h[0]$ after 128 iterations, which is done with the circular buffer; however, with the *A7 pointer we need to increment it by two bytes per interrupt (see Figure 5.19). This is done by the following assembly code:

```
LDH    *B6,A4     ; Get the sample from the serial port 0
NOP    4
STH    A4,*++A7   ; Pre-increment to offset x[n] and then store
                  ; the new sample in x_in array
```

After 128 loop iterations the output of the filter, acc1, should be:

$$\mathrm{acc1} = h_0 x_1 + h_{n-(N-1)} x_2 + \ldots + h_1 x_0$$

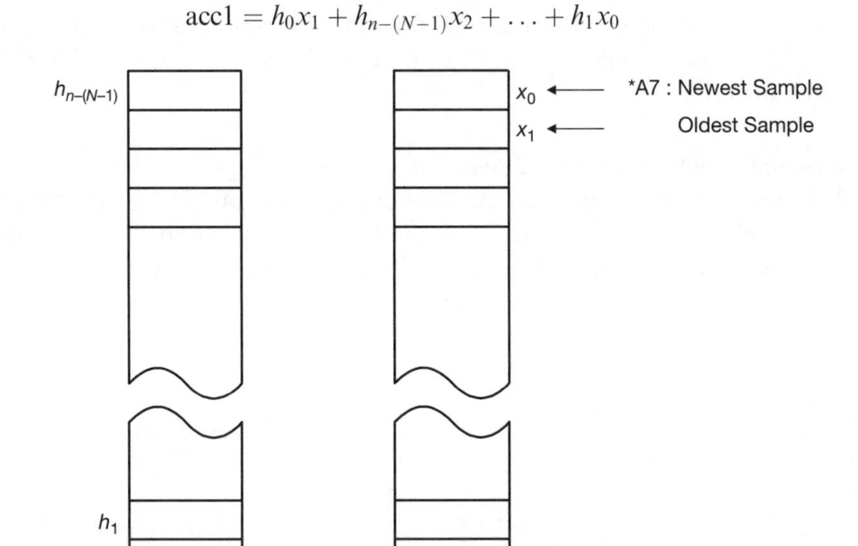

Figure 5.18 Initial condition of the circular buffers

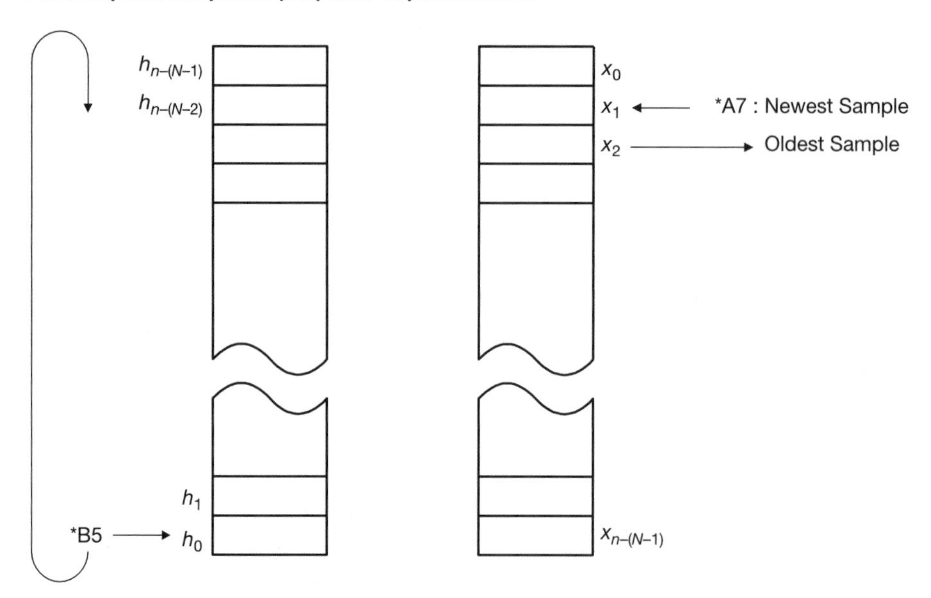

Figure 5.19 Circular buffer status after one complete iteration

5.3.4.3 Interfacing C and assembly

As a general rule, the code in C is used for initialisation and for non-critical (in terms of speed) code. In the following example, the C code is used for initialisation routines and the filter routine is written in assembly language. There are three different ways to interface the C and assembly codes. The first is for the C code to call the assembly function, the second is for an interrupt to call the assembly function, and the third is to call an assembly instruction using intrinsics.

Assembly function call by a C function

Programs 5.6 and 5.7 show an example of a C to assembly interface. Note that the compiler adds an underscore to the beginning of all labels, therefore the underscore has to be added to labels when using assembly. Programs 5.8 and 5.9 show the complete implementation.

```
// Program calling the fir_asm_init function
extern void fir_asm_init();
// use " far extern void fir_asm_init(); " if the function is located is at a distance
// greater than 1024k
// from the calling function which is main() in this case.
*
*
main() {
```

Program 5.6 C to assembly interface program

Program 5.6 continued

```
*
*
// call   fir_asm_init();
fir_asm_init();              // Before this function is called the return address is
                             // automatically saved in B3 register
for (;;);                    // Endless loop
       }
```

```
*assembly code containing the called function , _fir_asm_init.
    .sect  "iprog"
    .def   _fir_asm_init

_fir_asm_init
    MVKL        ; Do not use B3 register in this function unless it has been saved
*               ; previously or pushed on to the stack.
*
*
     B     B3  ; The calling function (C code) has saved the return address into B3
     NOP   5
```

Program 5.7 C to assembly interface called function

Assembly function call from a C function by an interrupt

This is very similar to the previous case except that the function has to be declared as an external interrupt, as follows:

```
extern interrupt void fir_asm (void);
```

Also the assembly function has to be hooked to the relevant interrupt, as follows:

```
intr_hook (fir_asm,CPU_INT15);
intr_map(CPU_INT15,ISN_RINT0);
```

Calling assembly functions by use of intrinsics

The ’C6x compiler allows the use of some functions called intrinsics that are identical to the assembly instruction. When using these intrinsics, it is possible to call an assembly language statement directly from the C code. The intrinsics are automatically inlined by the compiler. Refer to the *Optimizing C Compiler User’s Guide* (SPRU187) for more details. In this chapter, the `_mpy` intrinsic has been used.

The other possibility is to use the inline assembly language embedded in the C code as shown below:

```
asm ("MV     0x440,B2");
asm ("MVLKH  0x7,B2");
asm ("MVC    B2,AMR");
```

5.3.4.4 Interfacing C and assembly language for the FIR filter

Programs 5.8 and 5.9 are written in C and assembly language respectively in order to demonstrate the following points:

(1) The interfacing of C and assembly
(2) Use of circular addressing
(3) Implementation of an optimised FIR filter in assembly.

All the source and executable files can be found under the `F:\DSPCODE\Fir (Chap5)\ ASM\` directory.

```
/*------------------------------------------------------------------------------
    File:        fir_asm.c
    Description: Perform 128-tap fir filtering on codec input.
    Input:       Right channel of line-in
    Output:      Right channel = original signal
                 Left channel  = filtered signal
                 * Note: filter routine is optimised and linked separately.
------------------------------------------------------------------------------*/

#include <stdio.h>
#include <string.h>
#include <common.h>
#include <codec.h>
#include <mcbspdrv.h>
#include <board.h>
#include <stdlib.h>
#include <timer.h>
#include <intr.h>
#include <regs.h>

#define SAMPLING_RATE 44100

extern int inicodec (int sampling_rate);
extern far void fir_asm_init (void);
extern interrupt void fir_asm_intr();

#pragma DATA_ALIGN(h,256)      // Circular buffer of 128 short (2 bytes).
#pragma DATA_ALIGN(R_in,256)   // Circular buffer of 128 short (2 bytes).
short R_in[128];               // Input samples R_in[0] most recent, R_in[127] oldest.
short h[]=                     // Impulse response of FIR filter.
{
    -2,   10,   14,    7,    -7,   -17,   -13,     3,
    19,   21,    4,  -21,   -32,   -16,    18,    43,
    34,   -8,  -51,  -56,   -11,    53,    81,    41,
   -44, -104,  -81,   19,   119,   129,    24,  -119,
  -178,  -88,   95,  222,   171,   -41,  -248,  -266,
   -50,  244,  366,  181,  -195,  -457,  -353,    85,
   522,  568,  109, -540,  -831,  -424,   474,  1163,
   953, -245, -1661, -2042, -463,  2940,  6859,  9469,
  9469, 6859, 2940, -463, -2042, -1661,  -245,   953,
  1163,  474, -424, -831,  -540,   109,   568,   522,
```

Program 5.8 Control program for the FIR filter

Program 5.8 continued

```
    85,  -353,  -457,  -195,   181,   366,   244,   -50,
  -266,  -248,   -41,   171,   222,    95,   -88,  -178,
  -119,    24,   129,   119,    19,   -81,  -104,   -44,
    41,    81,    53,   -11,   -56,   -51,    -8,    34,
    43,    18,   -16,   -32,   -21,     4,    21,    19,
     3,   -13,   -17,    -7,     7,    14,    10,    -2
};

void Init_Interrupts ()
{
  intr_reset();                       // Reset the interrupt system,
                                      // disable all interrupts.

  INTR_CLR_FLAG (CPU_INT15);          // Clear previous interrupt request
  INTR_ENABLE (CPU_INT15);            // Enable cpu interrupt line 15
  intr_map(CPU_INT15,ISN_RINT0);      // Link receive interrupt of serial
                                      // port 0 to interrupt line 15.
  intr_hook (fir_asm_intr,CPU_INT15); // Assign the interrupt service routine
}

void main(void)
{
  evm_init();                         // Initialise EVM board.
  inicodec (SAMPLING_RATE);           // Initialise codec, adjust sampling rate.
  Init_Interrupts ();                 // Initialise interrupts and hook isr.
  fir_asm_init();
  INTR_GLOBAL_ENABLE();               // Enable global interrupt.

  /**************************************************
  Note: In this case, main loop must be empty,
        because the interrupt service routine
        uses some registers to implement circular
        buffers. Adding codes may damage the buffer.
  **************************************************/

  for (;;);                           // Main loop, does nothing.
}
```

```
        .def    _fir_asm_intr
        .def    _fir_asm_init
        .sect   "iprog"

        .ref    _h
        .ref    _R_in

_fir_asm_init
        MVKL    0x018C0000,B6
        MVKH    0x018C0000,B6

        MVK     0xFFFF0000,A10    ; Use A10 as a mask
        MVKH    0xFFFF0000,A10
        MVK     0x0000FFFF,A11    ; Use A11 as a mask
        MVKH    0x0000FFFF,A11
```

Program 5.9 Assembly code for the FIR filter

Program 5.9 continued

```
        MVKL    0x0440,B2         ; B5 and A7 used for circular addressing
        MVKLH   0x7,B2            ; N=7 -> 2^N+1 = 128 *2
        MVC     B2,AMR

        MVKL    _R_in,A7
        MVKH    _R_in,A7
        MVKL    _h,B5
        MVKH    _h,B5

        MVKL    127,B9
        MVKH    127,B9
        MV      B9,A9

        LDH     *++B5[b9],A4      ; Dummy read to point to h0.
        LDH     *++A7[a9],A4      ; Dummy read to point to R_in[128]
        NOP     4

        B       B3
        NOP     5

_fir_asm_intr

        ZERO    A5
        LDH     *B6,A4
        NOP     4
        MV      A4,A3             ; Preserve the input into A3
        AND     A3,A11,A3
        STH     A4,*++A7          ; Pre-increment to offset x[n]

        MVK     120,B0            ; Set the loop counter to 128 -8 = 120
        MVKH    120,B0

;*----------------------------------------------------------------------*
; PIPED LOOP PROLOG

          LDH     .D2T2   *B5++,B4
||        LDH     .D1T1   *A7++,A0

          LDH     .D2T2   *B5++,B4
||        LDH     .D1T1   *A7++,A0

   [ B0]  SUB     .L2     B0,0x1,B0
||        LDH     .D2T2   *B5++,B4
||        LDH     .D1T1   *A7++,A0

   [ B0]  B       .S2     loop
|| [ B0]  SUB     .L2     B0,0x1,B0
||        LDH     .D2T2   *B5++,B4
||        LDH     .D1T1   *A7++,A0

   [ B0]  B       .S2     loop
|| [ B0]  SUB     .L2     B0,0x1,B0
||        LDH     .D2T2   *B5++,B4
||        LDH     .D1T1   *A7++,A0

   [ B0]  B       .S2     loop
|| [ B0]  SUB     .L2     B0,0x1,B0
||        LDH     .D2T2   *B5++,B4
||        LDH     .D1T1   *A7++,A0
```

Program 5.9 continued

```
          MPY     .M1X    B4,A0,A4
|| [ B0]  B       .S2     loop
|| [ B0]  SUB     .L2     B0,0x1,B0
||        LDH     .D2T2   *B5++,B4
||        LDH     .D1T1   *A7++,A0

          MPY     .M1X    B4,A0,A4
|| [ B0]  B       .S2     loop
|| [ B0]  SUB     .L2     B0,0x1,B0
||        LDH     .D2T2   *B5++,B4
||        LDH     .D1T1   *A7++,A0
||        SHL     .S1     A4,1,A4
;** --------------------------------------------------------------------------*
loop:     ; PIPED LOOP KERNEL

          ADD     .L1     A4,A5,A5
||        MPY     .M1X    B4,A0,A4
|| [ B0]  B       .S2     loop
|| [ B0]  SUB     .L2     B0,0x1,B0
||        LDH     .D2T2   *B5++,B4
||        LDH     .D1T1   *A7++,A0
||        SHL     .S1     A4,1,A4
;** --------------------------------------------------------------------------*
L7:       ; PIPED LOOP EPILOGUE

          ADD     .L1     A4,A5,A5
||        MPY     .M1X    B4,A0,A4
||        SHL     .S1     A4,1,A4

          ADD     .L1     A4,A5,A5
||        MPY     .M1X    B4,A0,A4
||        SHL     .S1     A4,1,A4

          ADD     .L1     A4,A5,A5
||        MPY     .M1X    B4,A0,A4
||        SHL     .S1     A4,1,A4

          ADD     .L1     A4,A5,A5
||        MPY     .M1X    B4,A0,A4
||        SHL     .S1     A4,1,A4

          ADD     .L1     A4,A5,A5
||        MPY     .M1X    B4,A0,A4
||        SHL     .S1     A4,1,A4

          ADD     .L1     A4,A5,A5
||        SHL     .S1     A4,1,A4

          ADD     .L1     A4,A5,A5
||        SHL     .S1     A4,1,A4

          ADD     .L1     A4,A5,A5

          AND             A5,A10,A5
          OR              A5,A3,A5
          STW             A5,*B6[1] ; Output the input and the filtered signals

          NOP   1

          B       .S2     IRP
          NOP     5
```

Chapter 6

Infinite Impulse Response (IIR) filter implementation

6.1 Introduction

Infinite Impulse Response (IIR) filters are the first choice when speed is paramount and the phase non-linearity characteristic is acceptable. IIR filters are computationally more efficient than FIR filters since they require fewer coefficients due to the fact that they use feedback or poles. However, this feedback can result in the filter being unstable if the coefficients deviate from their true values. This can happen during coefficient scaling or quantisation. In this chapter it will be shown that the design and implementation of IIR filters are different from those for FIR filters.

The general equations of an IIR filter can be expressed as follows:

$$H(z) = \frac{b_0 + b_1 z^{-1} + \ldots + b_N z^{-N}}{1 + a_1 z^{-1} + \ldots + a_M z^{-M}} = \frac{\sum_{k=0}^{N} b_k z^{-k}}{1 + \sum_{k=1}^{M} a_k z^{-k}}$$

where a_k and b_k are the filter coefficients.

This transfer function can be factorised to give

$$H(z) = k\frac{(z - z_1)(z - z_2)\ldots(z - z_3)}{(z - p_1)(z - p_2)\ldots(z - p_3)} = \frac{Y(z)}{X(z)}$$

where:

$z_1, z_2, \ldots, z_N$ are the zeros,
$p_1, p_2, \ldots, p_N$ are the poles.

In terms of the difference equation that is useful for implementation, the transfer function leads to the following equation:

$$y(n) = \sum_{k=0}^{\infty} h(k)x(n-k)$$
$$= \sum_{k=0}^{N} b_k x(n-k) + \sum_{k=1}^{M} a_k y(n-k)$$

6.2 Design procedure

Similar to FIR design, there are five main steps to follow for designing an IIR filter:

(1) Filter specification (refer to Chapter 5)
(2) Coefficients calculation
(3) Appropriate structure selection
(4) Simulation (optional)
(5) Implementation.

6.3 Coefficients calculation

There are two main approaches for deriving the z-transfer function of an IIR filter. The first approach is based on direct placement of poles and zeros and the second is based on analogue filter design. Both approaches are described below.

6.3.1 Pole–zero placement approach

This is the easiest method for designing simple filters. All that is required is the knowledge that by placing a zero near to or on the unit circle in the z-plane, the transfer function will be minimised at these points, whereas by placing a pole near to or on the unit circle in the z-plane, the transfer function will be maximised at these points.

To obtain real coefficients, the poles and zeros must either be real or occur in complex conjugate pairs.

6.3.2 Analogue to digital filter design

This is the most popular method for calculating the filter coefficients. The popularity of this method comes from the rich analysis of the well-established analogue filters. There are two principal methods, as presented in this book; these are the bilinear transform (Section 6.3.3) and the impulse invariant (Section 6.3.4) methods.

6.3.3 Bilinear transform (BZT) method

The bilinear transform method is perhaps the most common and effective method for deriving a digital filter from its counterpart analogue filter. This method is relatively simple and consists of mapping the s-plane to the z-plane.

Since we know that $z = e^{j\omega T_s}$ and $s = j\omega$, we can establish the following relationship between the s and z domains:

$$z = e^{sT_s} \qquad [6.1]$$

or

$$s = \frac{1}{T_s} \ln z \qquad [6.2]$$

Substituting Equation [6.2] for s into a transfer function $H(s)$ will lead to an equation in $\ln(z)$ which is not practical. To eliminate the logarithmic term, an approximation to $\ln(z)$ is used:

$$\ln(z) = 2\left[\frac{z-1}{z+1} + \frac{1}{3}\left(\frac{z-1}{z+1}\right)^3 + \ldots + \frac{1}{n+1}\left(\frac{z-1}{z+1}\right)^{n+1} + \ldots\right] \qquad [6.3]$$

By considering only the first element in Equation [6.3], a new relationship between the s and z domains can be obtained, namely that:

$$s = \frac{1}{T_s}\ln(z) = \frac{2}{T_s}\frac{z-1}{z+1} \qquad [6.4]$$

As shown previously, Equation [6.4], which is known as the bilinear transform, is an approximation and therefore introduces distortion. However, Equation [6.4] offers a good approximation and is widely used. The bilinear z-transform mapping from the s- to the z-plane is shown in Figure 6.1.

The mapping from the s- to the z-plane introduces non-linearity between the analogue and digital frequencies as follows. By using

$$z = e^{j\omega_p T_s}$$

$$s = j\omega$$

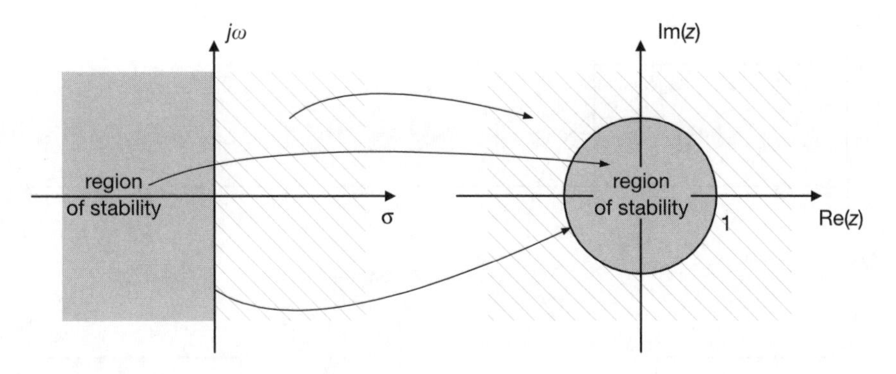

Figure 6.1 s- to z-plane mapping

and

$$s = \frac{2}{T_s} \frac{z-1}{z+1}$$

we can deduce the relationship between ω and ω_p to be

$$\omega = \frac{2}{T_s} \tan\left(\frac{\omega_p T_s}{2}\right) \qquad [6.5]$$

Equation [6.5] is sketched in Figure 6.2.

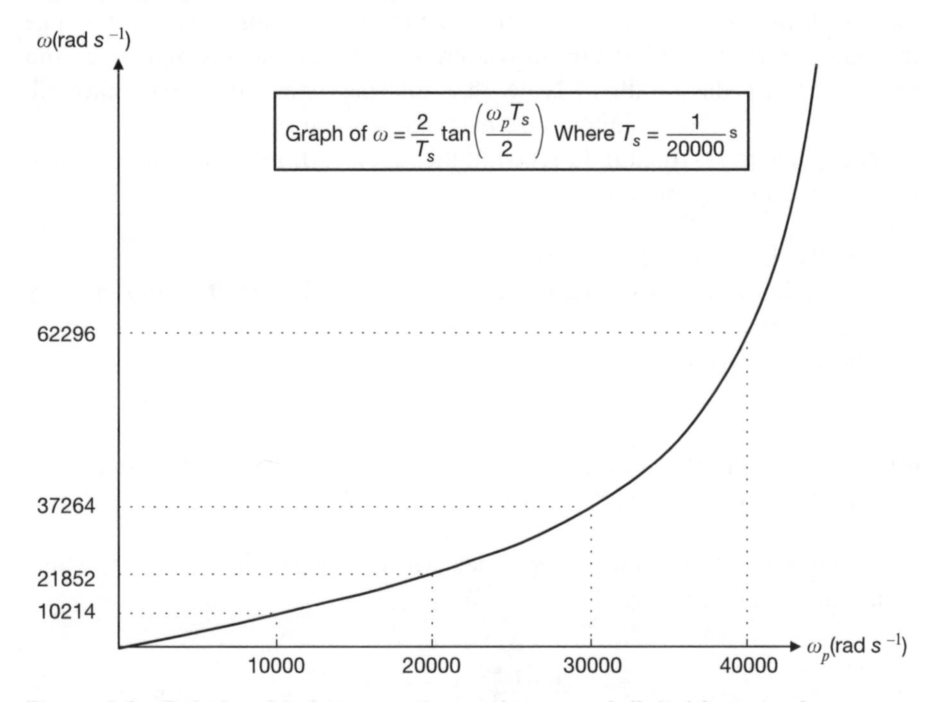

Figure 6.2 Relationship between the analogue and digital frequencies

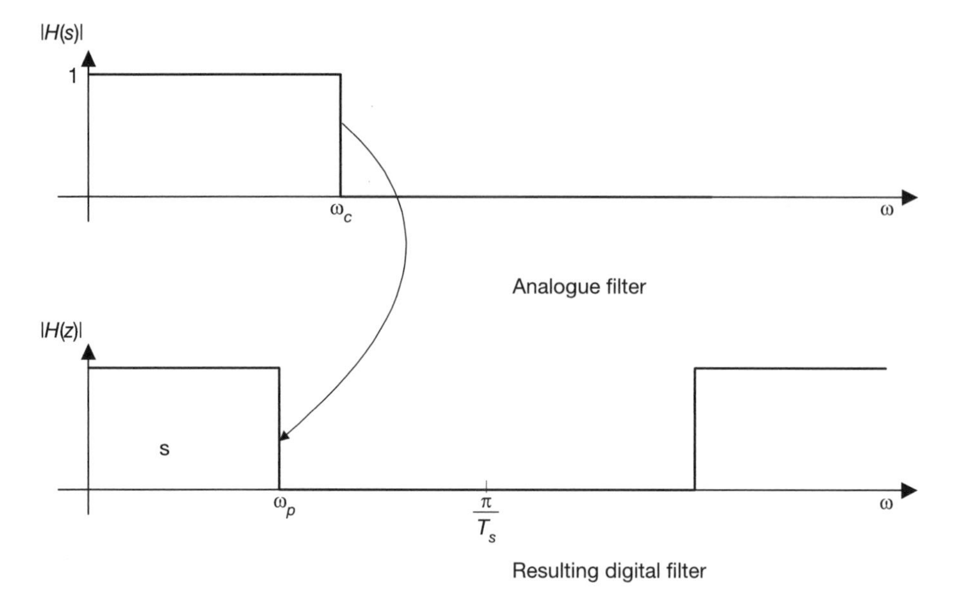

Figure 6.3 Relationship between the analogue and digital frequency responses when using the BZT

Equation [6.5] shows that when choosing a digital frequency, for instance the cut-off frequency, it has to be converted before being used as the analogue frequency. Figure 6.3 illustrates the difference between the analogue filter and its counterpart digital filter. Note that aliasing does not occur since all frequencies from zero to infinity are mapped between zero and π/T_s.

When deriving a digital filter with the bilinear transform the following logical design flow can be used.

(1) Specify the analogue filter prototype.
(2) Determine the cut-off frequency, ω_p, of the digital filter and find its equivalent analogue cut-off frequency, ω_c, using Equation [6.5]. This is known as 'pre-warping'.
(3) Denormalise the analogue filter by ω_c. This can be done by replacing s by s/ω_c.
(4) Finally, apply the bilinear transform to the filter obtained in step 3 by replacing s by $k(z-1)/(z+1)$ where $k = 2/T_s$.

To group steps (2)–(4) into a single step, let us express s in terms of z whilst taking them into account.

$$\omega_c = 2f_s \tan\left(\frac{\omega_p}{2f_s}\right) = k \tan\left(\frac{\omega_p}{2f_s}\right)$$

and

$$s = 2f_s \frac{z-1}{z+1} = k \frac{z-1}{z+1}$$

hence

$$\frac{s}{\omega_c} = \frac{z-1}{z+1} \tan^{-1}\left(\frac{\omega_p}{2f_s}\right)$$

which is equal to

$$\frac{z-1}{z+1} \tan^{-1}\left(\frac{\pi f_p}{f_s}\right) = \frac{1}{a} \frac{z-1}{z+1}$$

where

$$a = \tan\left(\frac{\pi f_p}{f_s}\right)$$

Note that k cancels out. Therefore,

$$H(s)|_{s=\frac{s}{\omega_c}} = H(s)|_{s=\frac{1}{\tan\left(\frac{\pi f_p}{f_s}\right)} \frac{z-1}{z+1}} = H(s)|_{s=\frac{1}{a}\frac{z-1}{z+1}}$$

Finally, we have two steps for converting an analogue filter to a digital filter using the BZT. These steps are:

(1) Specify the analogue filter prototype.
(2) Find $H(z)$ using

$$H(z) = H(s)|_{s=\frac{1}{\tan\left(\frac{\pi f_p}{f_s}\right)} \cdot \frac{z-1}{z+1}}$$

6.3.3.1 Practical example of the bilinear transform method

In this section the design of a digital filter to approximate a second-order lowpass analogue filter described by the following transfer function is considered:

$$H(s) = \frac{1}{s^2 + \sqrt{2}s + 1}$$

The digital filter should have a cut-off frequency of 6 kHz and a sampling frequency of 20 kHz.

6.3.3.2 Coefficients calculation

Now that the specifications have been given, we can proceed to step 2 to determine the $H(z)$ function.

$$H(z) = H(s)\Big|_{s=\frac{1}{\tan\left(\frac{\pi f_p}{f_s}\right)}\cdot\frac{z-1}{z+1}}$$

$$H(z) = \left.\frac{1}{s^2+\sqrt{2}s+1}\right|_{s=\frac{1}{a}\frac{z-1}{z+1}}$$

$$= \frac{1}{\frac{1}{a^2}\left(\frac{z-1}{z+1}\right)^2+\frac{\sqrt{2}}{a}\left(\frac{z-1}{z+1}\right)+1}$$

$$= \frac{a^2(z^2+2z+1)}{(z^2-2z+1)+\sqrt{2}a(z^2-1)+a^2(z^2+2z+1)}$$

$$= a^2\frac{z^2+2z+1}{z^2(1+\sqrt{2}a+a^2)+z(2a^2-2)+(1+a^2-\sqrt{2}a)}$$

$$H(z) = \frac{1+2z^{-1}+z^{-2}}{\left(\frac{1+\sqrt{2}a+a^2}{a^2}\right)+2\left(\frac{a^2-1}{a^2}\right)z^{-1}+\left(\frac{1+a^2-\sqrt{2}a}{a^2}\right)z^{-2}}$$

$$= \frac{b_0+b_1z^{-1}+b_2z^{-2}}{1+a_1z^{-1}+a_2z^{-2}}$$

where

$$a = \tan\left(\frac{\pi f_p}{f_s}\right) = \tan\left(\frac{\pi.6000}{20000}\right) = 1.376382$$

$$b_0 = \frac{a^2}{1+\sqrt{2}a+a^2}$$

$$b_1 = 2b_0$$

$$b_2 = b_0$$

$$a_1 = 2\frac{a^2-1}{(1+\sqrt{2}a+a^2)}$$

$$a_2 = \frac{1+a^2-\sqrt{2}a}{1+a^2+\sqrt{2}a}$$

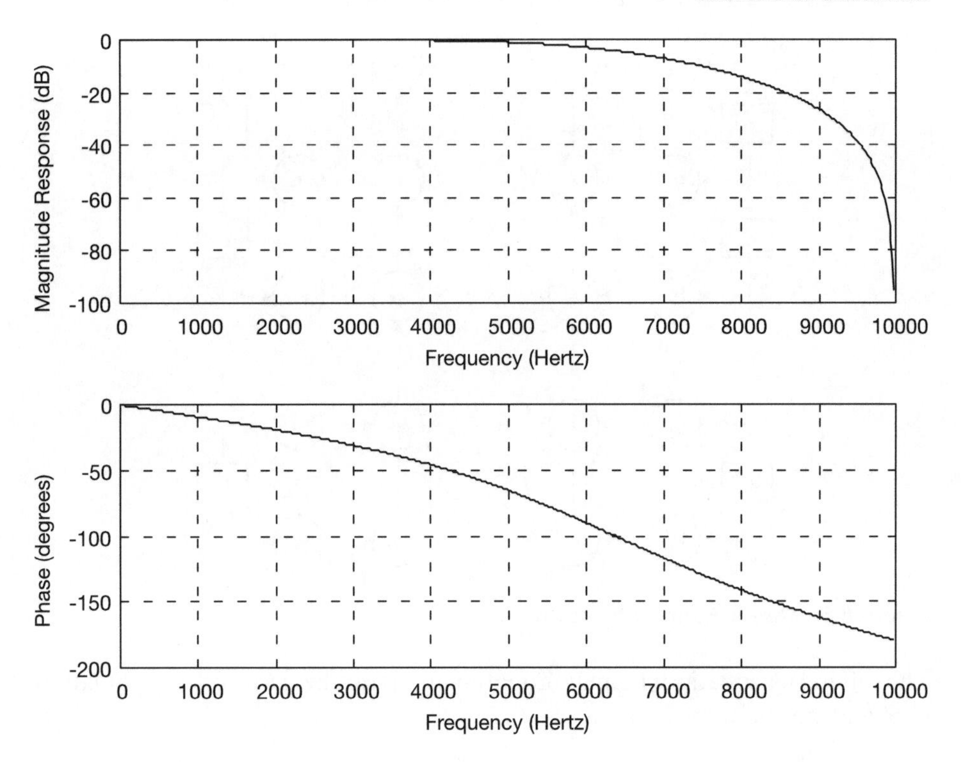

Figure 6.4 Transfer function of an IIR filter designed with the bilinear transform method

Figure 6.4 shows the transfer function of $H(z)$ using MATLAB.

6.3.3.3 Realisation structures

As stated in Section 5.3.3, Chapter 5, the choice of a structure depends on a few factors and can be obtained by manipulation of the transfer functions. In this chapter the two frequently used structures are described, that is the direct and cascade structures.

Direct form structure

As shown previously, the transfer function of an IIR filter can be expressed as shown in Equation [6.6]:

$$H(z) = \frac{Y(z)}{X(z)} = \frac{\sum_{n=0}^{N} b_n z^{-n}}{1 + \sum_{n=0}^{M} a_n z^{-n}} = \frac{b_0 + b_1 z^{-1} + \ldots + b_N z^{-N}}{1 + a_1 z^{-1} + \ldots + a_M z^{-M}} \qquad [6.6]$$

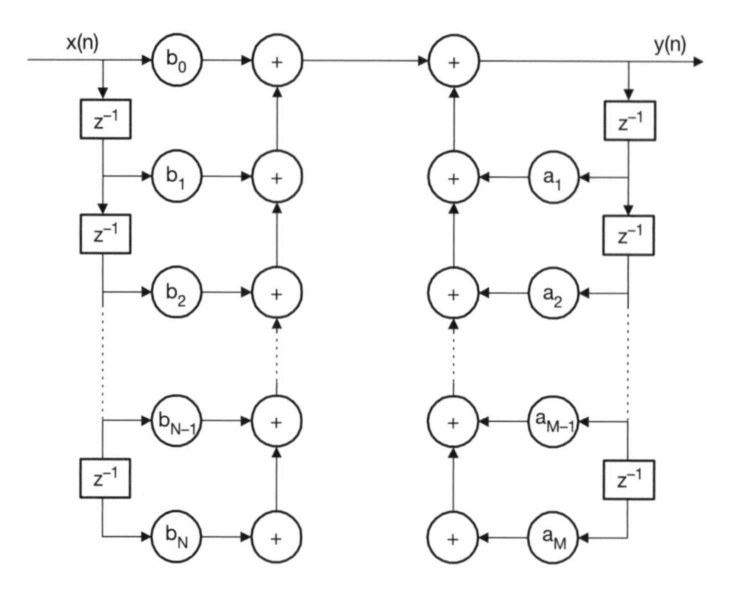

Figure 6.5 Direct Form I structure

The difference equation, namely Equation [6.7], can be derived from Equation [6.6] and can be written as:

$$y(n) = \sum_{k=0}^{N} b_k x(n-k) + \sum_{k=1}^{M} a_k y(n-k) \qquad [6.7]$$

Equation [6.7] leads to the direct structure shown in Figure 6.5 and is known as the Direct Form I structure. By using this structure $N + M$ delay elements are required. This also means that $N + M$ data memory moves are required; to reduce this number the transfer function can be manipulated as follows:

$$H(z) = H_1(z) \cdot H_2(z) = \frac{1}{1 + \sum_{k=1}^{N} a_k z^{-k}} \sum_{k=0}^{N} b_k z^{-k} \qquad (\text{for } N = M)$$

$$= \frac{P(z)}{X(z)} \cdot \frac{Y(z)}{P(z)}$$

where

$$\frac{P(z)}{X(z)} = \frac{1}{1 + \sum_{k=1}^{N} a_k z^{-k}}$$

implies that

$$P(z) = \frac{X(z)}{1 + \sum_{k=1}^{N} a_k z^{-k}}$$

and

$$\frac{Y(z)}{P(z)} = \sum_{k=0}^{N} b_k z^{-k}$$

implies that

$$Y(z) = P(z) \sum_{k=0}^{N} b_k z^{-k}$$

Taking the inverse z-transform of $P(z)$ and $Y(z)$ leads to Equations [6.8] and [6.9]:

$$p(n) = x(n) - \sum_{k=1}^{N} a_k p(n-k) \qquad [6.8]$$

$$y(n) = \sum_{k=0}^{N} b_k p(n-k) \qquad [6.9]$$

These lead to the realisation structure shown in Figure 6.6, which is known as the Direct Form II canonic realisation. Note that with this structure the

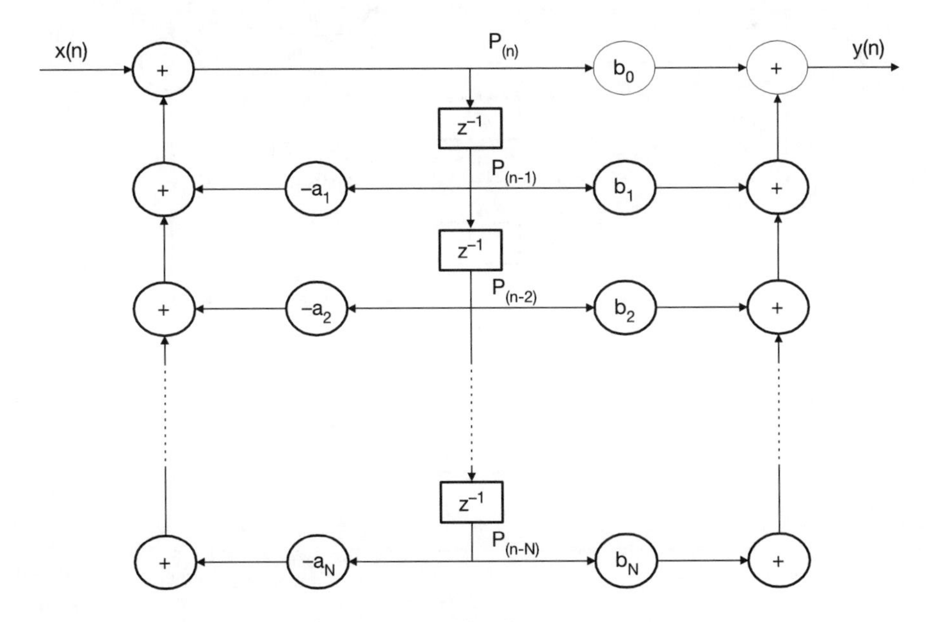

Figure 6.6 Direct Form II canonic realisation

number of delays has been reduced to N; this will save storage space and cycles during implementation.

Combining Equations [6.8] and [6.9] leads to

$$\begin{aligned} y(n) &= b_0 p(n) + \sum_{k=1}^{N} b_k p(n-k) \\ &= b_0 \left[x(n) - \sum_{k=1}^{N} a_k p(n-k) \right] + \sum_{k=1}^{N} b_k p(n-k) \\ &= b_0 x(n) + \sum_{k=1}^{N} (b_k - b_0 a_k) p(n-k) \end{aligned} \qquad [6.10]$$

Equation [6.10] leads to an alternative to the Direct Form II canonic realisation as shown in Figure 6.7. This structure has the advantage of scaling down the amplitude and therefore improves the performance of the filter.

Cascade realisation

To obtain the cascade realisation, the transfer function $H(z)$ is expressed as a product of transfer functions $H_1(z), H_2(z), \ldots, H_N(z)$:

$$H(z) = \frac{Y(z)}{X(z)} = H_1(z) \cdot H_2(z) \cdot \ldots \cdot H_N(z) \qquad [6.11]$$

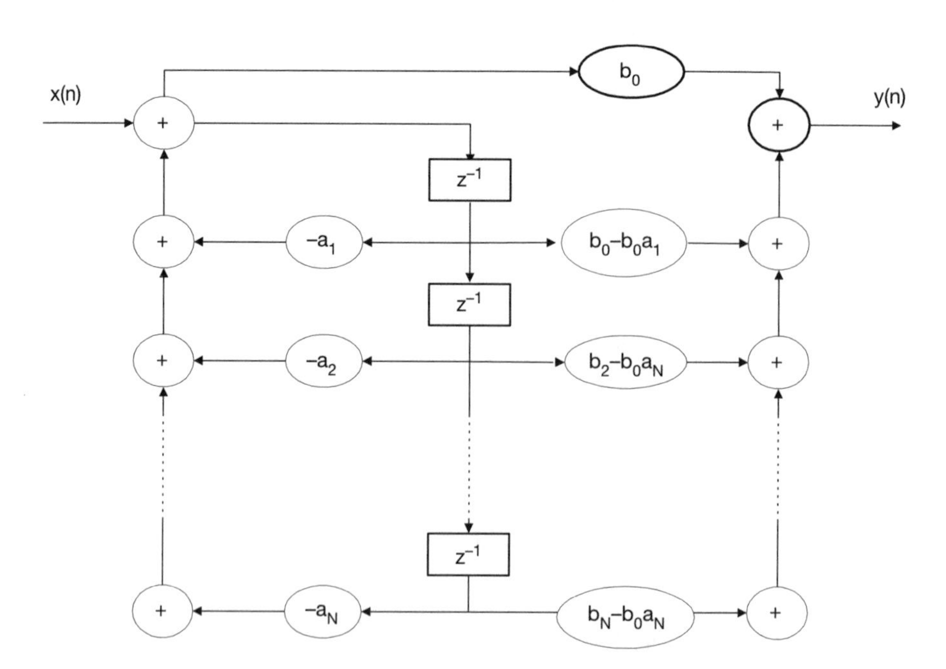

Figure 6.7 Alternative to the Direct Form II realisation

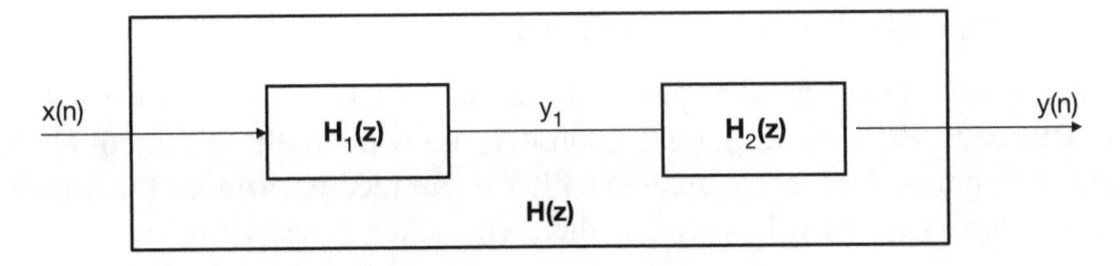

Figure 6.8 Cascade realisation using Direct Form II

$H_n(z)$ are often called bi-quadratic expressions, which can be determined by one of the methods shown above, and take the following form:

$$H_n(z) = \frac{b_{0n} + b_{1n}z^{-1} + b_{2n}z^{-2}}{1 + a_{1n}z^{-1} + a_{2n}z^{-2}}, \quad n = 1, 2, \ldots, N \qquad [6.12]$$

$H_n(z)$ can be implemented with the Direct Form I or II as shown above. However, the advantage of the Direct Form II structure makes it the first choice. Let us take a two-stage cascade realisation using the Direct Form II structure. The overall representation of $H(z)$ is shown in Figure 6.8. Equations [6.8] and [6.9] are rewritten to take into account the cascade realisation. This is expressed in Equations [6.13] to [6.19].

Equations for $H_1(z)$:

$$p_1(n) = x(n) - a_{11}p_1(n-1) - a_{21}p_1(n-2) \qquad [6.13]$$

$$y_1(n) = b_{01}p_1(n) + b_{11}p_1(n-1) + b_{21}p_1(n-2) \qquad [6.14]$$

Equations for $H_2(z)$:

$$p_2(n) = y_1(n) - a_{12}p_2(n-1) - a_{22}p_2(n-2) \qquad [6.15]$$

$$y(n) = b_{02}p_2(n) + b_{12}p_2(n-1) + b_{22}p_2(n-2) \qquad [6.16]$$

Substituting $y_1(n)$ from Equation [6.14] into Equation [6.15] leads to

$$p_1(n) = x(n) - a_{11}p_1(n-1) - a_{21}p_1(n-2) \qquad [6.17]$$

$$p_2(n) = b_{01}p_1(n) + b_{11}p_1(n-1) + b_{21}p_1(n-2) - a_{12}p_2(n-1) - a_{22}p_2(n-2) \qquad [6.18]$$

$$y(n) = b_{02}p_2(n) + b_{12}p_2(n-1) + b_{22}p_2(n-2) \qquad [6.19]$$

Equations [6.17], [6.18] and [6.19] show that in order to implement this filter, only the coefficients p_n are required and that there is no need to be concerned by the intermediate outputs y_n.

6.3.4 Impulse invariant method

It has been shown in the previous chapter that a filter can be fully characterised when its impulse response is known. With this method the impulse response, $h(t)$, of an analogue filter is sampled to produce the impulse response $h(nT)$ which will represent the digital filter as shown below:

$$h(n) = h(t)|_{t=nT} = h(nT) \qquad [6.20]$$

Given an analogue filter as expressed in Equation [6.21],

$$H(s) = \frac{(s - z_1)(s - z_2)\cdots(s - z_n)}{(s - p_1)(s - p_2)\cdots(s - p_n)} = \frac{b(s)}{a(s)} \qquad [6.21]$$

where z_n and p_n are the zeros and poles respectively, the impulse of this filter will be given by Equation [6.22]:

$$h(t) = L^{-1}[H(s)] \qquad [6.22]$$

where L^{-1} is the inverse Laplace transform.

To calculate the inverse transform of $H(s)$, the first step is to expand $H(s)$ in terms of partial fractions as shown in Equation [6.23]:

$$H(s) = \sum_{i=1}^{N} \frac{r_i}{s - p_i} \qquad [6.23]$$

The r_i, $i = 1, \ldots, N$, are known as the residues and can be calculated one by one by multiplying the following equation by the appropriate term as shown below:

$$H(s) = \frac{b(s)}{a(s)} = \frac{r_1}{s - p_1} + \frac{r_2}{s - p_2} + \ldots + \frac{r_n}{s - p_N} + k \qquad [6.24]$$

For the calculation of r_i, Equations [6.23] and [6.24] lead to

$$\frac{b(s)(s - p_1)}{a(s)} = r_1 + \frac{r_2(s - p_1)}{s - p_2} + \frac{r_3(s - p_1)}{s - p_3} + \ldots + \frac{r_n(s - p_1)}{s - p_N}$$

$$= r_1 + (s - p_1)\left[\frac{r_2}{s - p_2} + \frac{r_3}{s - p_3} + \ldots + \frac{r_N}{s - p_N}\right]$$

Therefore

$$\left.\frac{b(s)(s - p_1)}{a(s)}\right|_{s=p_1} = \left.\frac{b(s)}{(s - p_2)(s - p_3)\cdots(s - p_N)}\right|_{s=p_1} = r_1$$

Hence

$$r_1 = \left. \frac{b(s)}{(s-p_2)(s-p_3)\cdots(s-p_N)} \right|_{s=p_1}$$

The other residues can be calculated in a similar way.

Now that we have all the residues, we can easily find the inverse Laplace transform of $H(s)$ by using the following:

$$L^{-1}\left(\frac{r_i}{s-p_i}\right) = r_i e^{p_i t}$$

Therefore:

$$L^{-1}(H(s)) = h(t) = \sum_{i=1}^{N} r_i e^{p_i t}$$

By sampling the impulse response we obtain

$$h[nT] = \sum_{i=1}^{N} r_i e^{p_i nT} = \sum_{i=1}^{N} r_i (e^{p_i T})^n$$

Therefore

$$H(z) = \sum_{n=0}^{\infty} h(nT) z^{-n} = \sum_{n=0}^{\infty} \left[\sum_{i=1}^{N} r_i (e^{p_i T})^n \right] z^{-n}$$

From z-transform tables we know that

$$\sum_{n=0}^{\infty} e^{pnT} z^{-n} = \frac{z}{z - e^{pT}}$$

and so

$$H(z) = \sum_{i=1}^{N} \frac{r_i z}{z - e^{p_i T}}$$

In conclusion, in order to design a digital filter by the impulse invariant method the following steps are required:

(1) Specify the analogue prototype.
(2) Denormalise the filter by replacing s by s/ω_c, where ω_c is the cut-off frequency.
(3) Expand $H(s)$ in terms of partial fractions.

6.3.4.1 Practical example of the impulse invariant method

Let us design a lowpass digital filter which has a cut-off frequency of 2 kHz and a sampling frequency of 20 kHz and is based on an analogue filter with the following transfer function, $H(s)$:

$$H(s) = \frac{1}{s^2 + \sqrt{2}s + 1}$$

The solution is:

$$H(s) = H(s)|_{s=\frac{s}{\omega_c}} = \frac{{\omega_c}^2}{s^2 + \sqrt{2}\omega_c s + {\omega_c}^2}$$

$$= \frac{{\omega_c}^2}{(s - p_1)(s - p_2)} = \frac{r_1}{s - p_1} + \frac{r_2}{s - p_2}$$

where:

$$p_{1,2} = \frac{-\sqrt{2}\omega_c \pm \sqrt{-2{\omega_c}^2}}{2}$$

$$p_1 = \frac{-\sqrt{2}\omega_c - j\sqrt{2}\omega_c}{2} = \frac{-\omega_c\sqrt{2}(1 + j)}{2}$$

and

$$p_2 = \frac{-\omega_c\sqrt{2}(1 - j)}{2} = {p_1}^*$$

Let us now find r_1 and r_2:

$$\frac{{\omega_c}^2}{(s - p_1)(s - p_2)}(s - p_1) = r_1 + \frac{r_2}{s - p_2}(s - p_1)$$

Hence

$$r_1 = \left.\frac{{\omega_c}^2}{(s - p_2)}\right|_{s=p_1} = \frac{{\omega_c}^2}{p_1 - p_2}$$

$$= \frac{{\omega_c}^2}{p_1 - {p_1}^*} = -j\frac{\omega_c}{\sqrt{2}}$$

Similarly

$$r_2 = \left.\frac{{\omega_c}^2}{(s - p_1)}\right|_{s=p_2} = \frac{{\omega_c}^2}{p_2 - p_1}$$

$$= \frac{{\omega_c}^2}{p_2 - {p_2}^*} = +j\frac{\omega_c}{\sqrt{2}}$$

$$= j\frac{\omega_c}{\sqrt{2}} = r_1^*$$

Therefore

$$\begin{aligned} H(z) &= \frac{r_1 z}{z - e^{p_1 T}} + \frac{r_2 z}{z - e^{p_2 T}} \\ &= \frac{r_1}{1 - e^{p_1 T} z^{-1}} + \frac{r_2}{1 - e^{p_2 T} z^{-1}} \\ &= \frac{r_1(1 - e^{p_2 T} z^{-1}) + r_2(1 - e^{p_1 T} z^{-1})}{(1 - e^{p_1 T} z^{-1})(1 - e^{p_2 T} z^{-1})} \\ &= \frac{r_1 + r_2 - (r_1 e^{p_1 T} + r_2 e^{p_2 T}) z^{-1}}{1 - (e^{p_1 T} + e^{p_2 T}) z^{-1} + e^{(p_1 + p_2)T} z^{-2}} \end{aligned} \quad [6.25]$$

Knowing that

$$r_1 = r_2^* = r_r + jr_i$$

$$p_1 = p_2^* = p_r + jp_i$$

it follows that

$$\begin{aligned} e^{p_1 T} + e^{p_2 T} &= e^{(p_r + jp_i)T} + e^{(p_r - jp_i)T} \quad [6.26] \\ &= 2e^{p_r T} \cos(p_i T) \end{aligned}$$

$$e^{(p_1 + p_2)T} = e^{2p_r T} \quad [6.27]$$

and

$$\begin{aligned} r_1 e^{p_1 T} + r_2 e^{p_2 T} &= (r_r + jr_i)(e^{p_r T + jp_i T}) + (r_r - jr_i)(e^{p_r T - jp_i T}) \\ &= (r_r + jr_i)(e^{+p_r T})(e^{+jp_i T}) + (r_r - jr_i)(e^{+p_r T})(e^{-jp_i T}) \\ &= r_r(e^{+p_r T}(e^{+jp_i T} + e^{-jp_i T})) + jr_i(e^{+p_r T}(e^{+jp_i T} - e^{-jp_i T})) \\ &= r_r e^{+p_r T}(2\cos(p_i T)) + jr_i e^{+p_r T}(-2j \sin(p_i T)) \\ &= [2r_r \cos(p_i T) + 2r_i \sin(p_i T)] e^{+p_r T} \end{aligned} \quad [6.28]$$

By substituting Equations [6.26]–[6.28] into [6.25] we obtain

$$\begin{aligned} H(z) &= \frac{2r_r - 2[r_r \cos(p_i T) + r_i \sin(p_i T)] e^{+p_r T} z^{-1}}{1 - 2e^{p_r T} \cos(p_i T) z^{-1} + e^{2p_r T} z^{-2}} \\ &= \frac{b_0 + b_1 z^{-1}}{1 + a_1 z^{-1} + a_2 z^{-2}} \end{aligned}$$

where

$$r_r = 0$$

$$r_i = -\frac{\omega_c}{\sqrt{2}}$$

$$p_i = p_r = -\frac{\omega_c\sqrt{2}}{2}$$

$$b_0 = 2r_r = 0$$

$$b_1 = 2[r_r \cos(p_i T) + r_i \sin(p_i T)]e^{p_r T}$$

$$= 2[r_i \sin(p_i T)]e^{p_r T} = 2\left[j\frac{\omega_c}{2}\sin\left(-\frac{\omega_c\sqrt{2}}{2}T\right)\right]e^{-\left(\frac{\omega_c\sqrt{2}}{2}T\right)}$$

$$a_1 = 2e^{p_r T_s}\cos(p_i T) = 2\left[j\frac{\omega_c}{2}\cos\left(-\frac{\omega_c\sqrt{2}}{2}T\right)\right]e^{-\left(\frac{\omega_c\sqrt{2}}{2}T\right)}$$

$$a_2 = e^{2p_r T} = e^{-2\left(\frac{\omega_c\sqrt{2}}{2}T\right)}$$

When the sampling frequency is 20 kHz $(1/T)$ and the cut-off frequency is 2 kHz, the coefficients are as follows:

$$b_0 = 0$$

$$b_1 = 4.8984\text{e} + 003$$

$$a_1 = -1.1580$$

$$a_2 = 0.4112$$

By using MATLAB one can verify these results by simply using the following commands:

```
T=1/20000;
Wc=2*pi*2000;
Rr=0;
ri=-wc/sqrt(2);
pii=-wc/sqrt(2);
pr=pii;
b0=2*rr
b1=2*(ri*sin(pii*T))*exp(pr*T)
a1=-2*exp(pr*T)*cos(pii*T)
a2=exp(2*pr*T)
end
```

The output of the above program is shown in Figure 6.9. The laborious task of finding the coefficients can be overcome by using the `impinvar` MATLAB command.

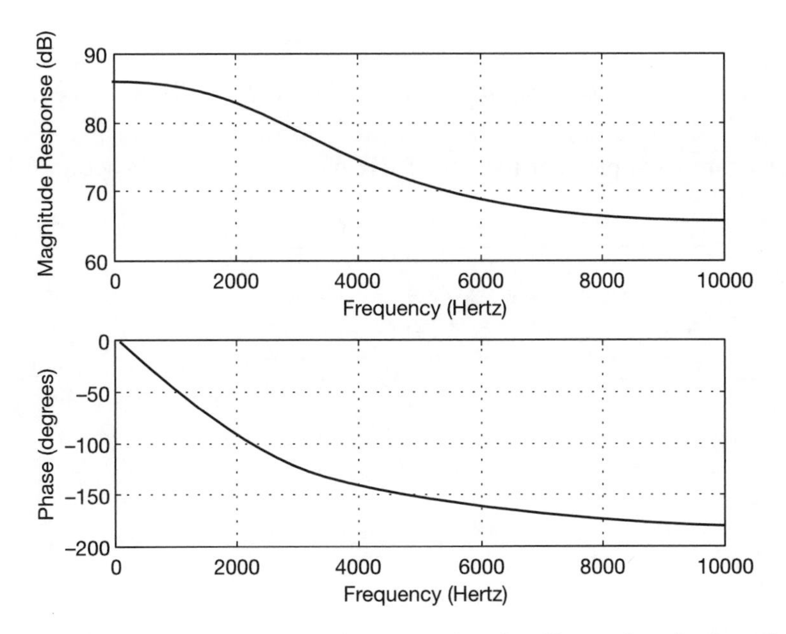

Figure 6.9 Frequency response of a second-order filter using the impulse invariant method

6.4 IIR filter implementation

Now that the transfer function has been determined, by choosing a realisation structure, the digital filter can finally be implemented. In this example the Direct Form II is chosen, using the coefficients calculated in Section 6.3.3.2; this is shown in Figure 6.10, and Equations [6.8] and [6.9] can be reduced to Equations [6.29] and [6.30]:

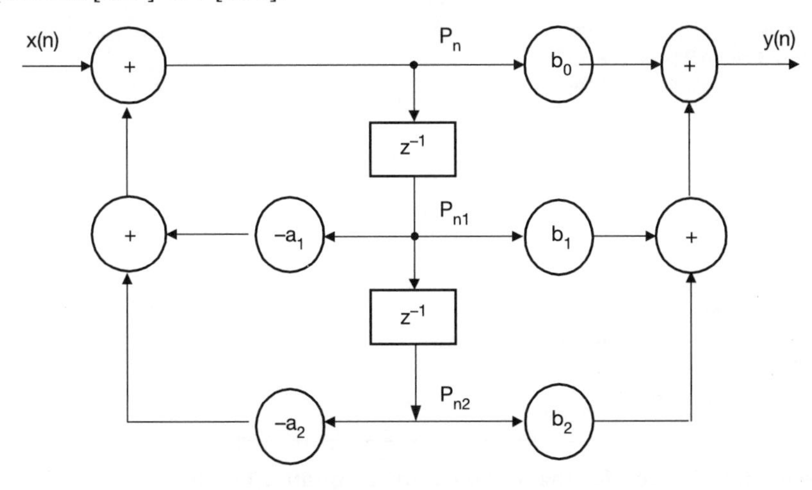

Figure 6.10 Direct Form II structure

$$p(n) = x(n) - a_1 p(n-1) - a_2 p(n-2) \quad [6.29]$$

$$y(n) = b_0 p(n) - b_1 p(n-1) - b_2 p(n-2) \quad [6.30]$$

The program to implement the above IIR filter is shown in Program 6.1.

```
/*---------------------------------------------------------------------
 File:        irr.c
 Description: Perform 1-stage and 4-stage quad IIR filter and output
              to left and right channel of line-out respectively.
              * Note: cut-off of single stage is 6 kHz @ 20 kHz.
 ---------------------------------------------------------------------*/

#include <stdio.h>
#include <string.h>
#include <common.h>
#include <codec.h>
#include <mcbspdrv.h>
#include <board.h>
#include <stdlib.h>
#include <timer.h>
#include <intr.h>
#include <regs.h>

#define SAMPLING_RATE 20000

extern int inicodec (int sampling_rate);

#pragma CODE_SECTION(isr_rint0,".iprog")

interrupt void isr_rint0 (void)
{
  short a1 = 0xd0b4;    // Negative value of a1
  short a2 = 0xe6f0;    // Negative value of a2

  short b0 = 0x3217;
  short b1 = 0x642e;
  short b2  = 0x3217;
  static short d01=0, d02=0, d00;
  static short d11=0, d12=0, d10;
  static short d21=0, d22=0, d20;
  static short d31=0, d32=0, d30;
  short xn, y0, y1, y2, y3;
  int prod1, prod2, prod3, prod4, prod5;
  int sin, sout;

  //============= pre filter =============
  sin = MCBSP0_DRR;
  //============= stage 0 =============
  xn = (short) (sin & 0x000ffff);
  prod1 = _mpy(d02,a2)>>15;
  prod2 = _mpy(d01,a1)>>15;
  d00 = xn + (short)(prod1 + prod2);
```

Program 6.1 C code for the implementation of an IIR filter

Program 6.1 continued

```
  prod3 = _mpy(d01,b1);
  prod4 = _mpy(d02,b2);
  prod5 = _mpy(d00,b0);
  y0 = (short)((prod3+prod4+prod5)>>15);
  d02 = d01;
  d01 = d00;
  //============= stage 1 =============
  prod1 = _mpy(d12,a2)>>15;
  prod2 = _mpy(d11,a1)>>15;
  d10 = y0 + (short)(prod1 + prod2);
  prod3 = _mpy(d11,b1);
  prod4 = _mpy(d12,b2);
  prod5 = _mpy(d10,b0);
  y1 = (short)((prod3+prod4+prod5)>>15);
  d12 = d11;
  d11 = d10;
  //============= stage 2 =============
  prod1 = _mpy(d22,a2)>>15;
  prod2 = _mpy(d21,a1)>>15;
  d20 = y1 + (short)(prod1 + prod2);
  prod3 = _mpy(d21,b1);
  prod4 = _mpy(d22,b2);
  prod5 = _mpy(d20,b0);
  y2 = (short)((prod3+prod4+prod5)>>15);
  d22 = d21;
  d21 = d20;
  //============= stage 3 =============
  prod1 = _mpy(d32,a2)>>15;
  prod2 = _mpy(d31,a1)>>15;
  d30 = y2 + (short)(prod1 + prod2);
  prod3 = _mpy(d31,b1);
  prod4 = _mpy(d32,b2);
  prod5 = _mpy(d30,b0);
  y3 = (short)((prod3+prod4+prod5)>>15);
  d32 = d31;
  d31 = d30;
  //============= post filter =============
  sout = (((int)y3)<<16)|(((int)y0)&0xffff);
  MCBSP0_DXR = sout;

  return;
}

void Init_Interrupts ()
{
  intr_reset();                     // Reset the interrupt system,
                                    // disable all interrupts.
  INTR_CLR_FLAG (CPU_INT15);        // Clear previous interrupt request
  INTR_ENABLE (CPU_INT15);          // Enable cpu interrupt line 15
  intr_map(CPU_INT15,ISN_RINT0);    // Link receive interrupt of serial
                                    // port 0 to interrupt line 15.
  intr_hook (isr_rint0,CPU_INT15);  // Assign the interrupt service routine
}
```

Program 6.1 continued

```
void main(void)
{
  evm_init();                       // Initialise EVM board.
  inicodec (SAMPLING_RATE);         // Initialise codec, adjust sampling rate.
  Init_Interrupts ();               // Initialise interrupts and hook isr.
  INTR_GLOBAL_ENABLE();             // Enable global interrupt.

  for (;;);                         // Main loop, does nothing.
}
```

Note that a cascade of three second-order filters was implemented. The left output channel of the codec represents the output of the first-stage filter and the right channel represents the output of the last filter stage. The filter was also implemented in linear assembly. Programs 6.2 and 6.3 show how to interface C and linear assembly in order to pass parameters and implement an IIR filter; the programs are self-explanatory. All the programs can be found in the `F:\DSPCODE\Iir (Chap6)\C` or `F:\DSPCODE\Iir (Chap6)\SA\` directories.

```
/*-----------------------------------------------------------------------
 File:        irr_sa.c
 Description: Perform 1-stage and 4-stage quad IIR filter and output
              to left and right channel of line-out respectively.
              * Note: cut-off of single stage is 6 kHz @ 20 kHz.
*************************************************************************/
#include <stdio.h>
#include <string.h>
#include <common.h>
#include <codec.h>
#include <mcbspdrv.h>
#include <board.h>
#include <stdlib.h>
#include <timer.h>
#include <intr.h>
#include <regs.h>

#define SAMPLING_RATE 20000

extern int inicodec (int sampling_rate);

#pragma CODE_SECTION(isr_rint0,".iprog")

unsigned int x_ptr = 0x18c0000;
unsigned int y_ptr = 0x18c0004;

short delays[2]={0x0,0x0};
int mask = 0x0000ffff;
short an1 = 0xd0b4;
short an2 = 0xe6fe;
short bn0 = 0x3217;
short bn1 = 0x642e;
short bn2 = 0x3217;
```

Program 6.2 Main program

Program 6.2 continued

```
interrupt void isr_rint0 (void)
{
  iir_sa(an1,an2,bn0,bn1,bn2,delays,x_ptr,y_ptr,mask);

  return;
}

void Init_Interrupts ()
{
  intr_reset();                    // Reset the interrupt system,
                                   // disable all interrupts.
  INTR_CLR_FLAG (CPU_INT15);       // Clear previous interrupt request
  INTR_ENABLE (CPU_INT15);         // Enable cpu interrupt line 15
  intr_map(CPU_INT15,ISN_RINT0);   // Link receive interrupt of serial
                                   // port 0 to interrupt line 15.
  intr_hook (isr_rint0,CPU_INT15); // Assign the interrupt service routine
}

void main(void)
{
  evm_init();                      // Initialise EVM board.
  inicodec (SAMPLING_RATE);        // Initialise codec, adjust sampling rate.
  Init_Interrupts ();              // Initialise interrupts and hook isr.
  INTR_GLOBAL_ENABLE();            // Enable global interrupt.

  for (;;);                        // Main loop, does nothing.
}
```

```
         .def      _iir_sa

         .sect     ".iprog"

_iir_sa  .cproc    an1, an2, bn0, bn1, bn2, delays, x_ptr, y_ptr,mask

         .reg      p0, p1, p2

         .reg      prod1, prod2, prod3, prod4, prod5
         .reg      sum1, sum2, sum3
         .reg      x, ref, y0,y1

         .reg      p0, p1, p2

         LDW       *x_ptr, x
         SHR       x, 16, x
         SHL       x, 16, ref

;============= Stage 0 ================

         LDH       *+delays[0], p1
         LDH       *+delays[1], p2

         MPY       an1, p1, prod1
         MPY       an2, p2, prod2
         ADD       prod1, prod2, sum1
         SHR       sum1, 15, sum1
         ADD       x, sum1, p0
```

Program 6.3 IIR filter written in linear assembly

Program 6.3 continued

```
        MPY         bn0, p0, prod3
        MPY         bn1, p1, prod4
        MPY         bn2, p2, prod5
        ADD         prod4, prod5, sum2
        ADD         prod3, sum2, sum3
        SHRu        sum3, 15, y0

        STH         p1, *+delays[1]
        STH         p0, *+delays[0]

        AND         y0, mask, y0
        OR          ref, y0, y0
        STW         y0, *y_ptr
        add         an1,y0,y0
        add         bn0,y0,y0
        add         bn2,y0,y0
        add         bn1,y0,y0
        add         an2,y0,y0
        .return     y0
        .endproc
```

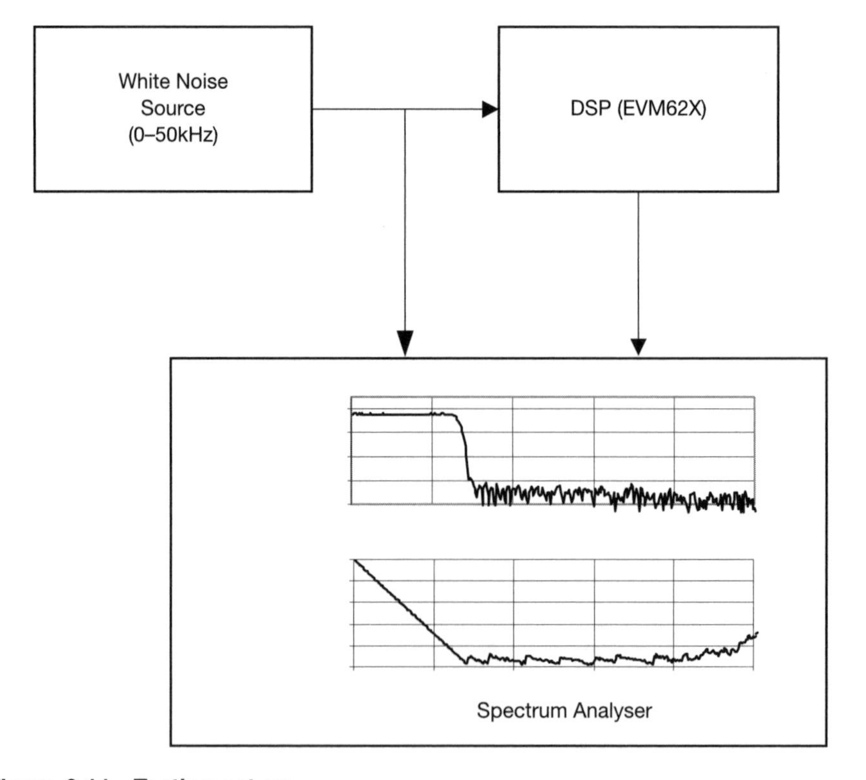

Figure 6.11 Testing set-up

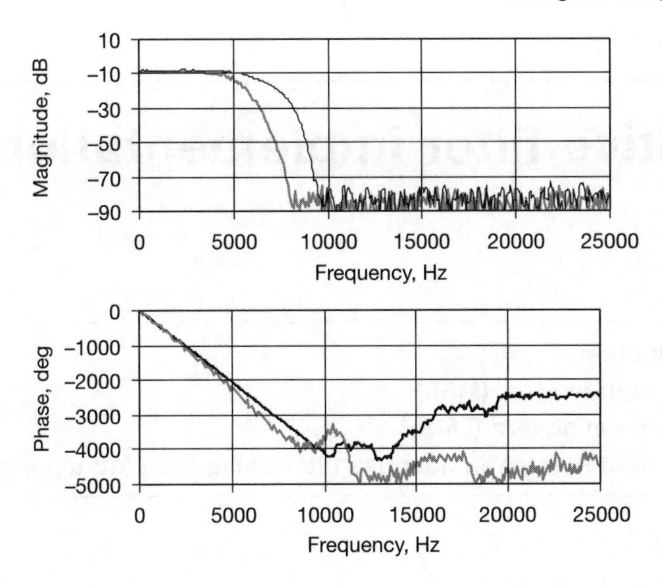

Figure 6.12 Transfer function of the designed IIR filter

6.5 Testing the designed IIR filter

The set-up shown in Figure 6.11 is used to measure the transfer function which is shown in Figure 6.12. Figure 6.12 clearly shows that the theoretical transfer function matches well with that of the implemented filter shown in Figure 6.4. The darker traces refer to the outputs of the first stage filter (right channel) and lighter traces refer to the last stage of the filter (left channel).

Chapter 7

Adaptive filter implementation

7.1 Introduction

Adaptive filters differ from other filters such as the FIR and IIR in the sense that the filter coefficients are not fixed and determined by some desired specifications, as shown in Chapters 5 and 6. With adaptive filters the specifications are not known and change with time. Applications of adaptive filters are numerous and include process control, medical instrumentation, speech processing, echo and noise cancellation and channel equalisation.

The general procedure of constructing an adaptive filter is first to choose an FIR or IIR filter, and, secondly, to have a mechanism or algorithm to optimally adjust the FIR or IIR coefficients (see Figure 7.1) (Widrow and Stearns, 1985; Bozic, 1994; Haykin, 1996).

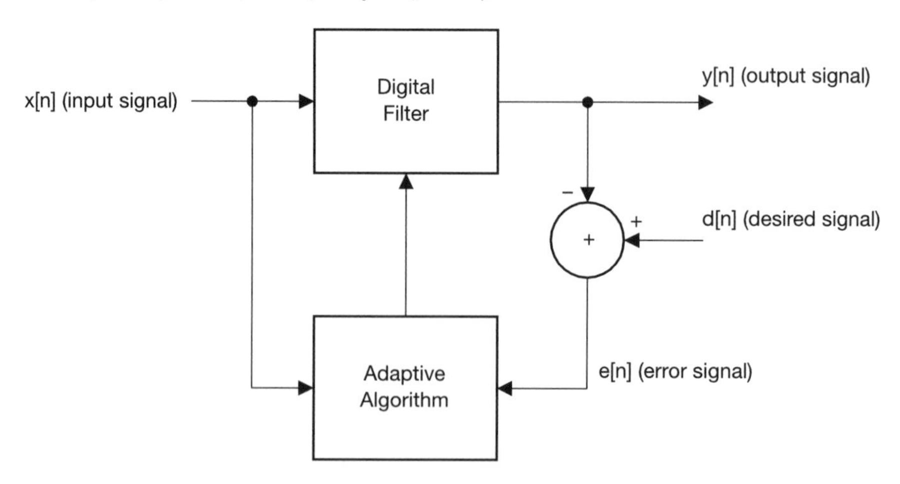

Figure 7.1 Basic block diagram of an adaptive filter

The real challenge for designing an adaptive filter resides with the adaptive algorithm. The latter needs to be practical to implement, adapt the filter coefficients quickly, and provide the desired performance.

How is the performance measured? The main criterion that provides a good measure of the performance is based on the mean square error (MSE).

This chapter will show how to calculate the filter coefficients using the MSE criterion before presenting the least mean square (LMS) algorithm. Finally, it will show how the LMS algorithm can be implemented in both C and Assembly.

7.2 Mean square error (MSE)

Referring to Figure 7.1 we can write the following equations:

$$y(n) = \sum_{k=0}^{N-1} h(k)x(n-k) \qquad [7.1]$$

$$e(n) = d(n) - y(n) = d(n) - \sum_{k=0}^{N-1} h(k)x(n-k) \qquad [7.2]$$

Let E[.] be the expectation operator. Taking the expectation of $e^2(n)$:

$$\begin{aligned} E[e^2(n)] &= E[(d(n) - y(n))^2] \\ &= E[d^2(n)] - 2E[d(n)y(n)] + E[y^2(n)] \end{aligned}$$

Using the following vector notation

$$y(n) = \sum_{i=0}^{N-1} h_i x(n-i)$$

can be written as

$$y(n) = H^T X(n)$$

The expectation of $e^2(n)$ is now

$$E[e^2(n)] = E[d^2(n)] - 2E[d(n)X^T(n)]H + E[H^T X(n) X^T(n) H]$$

Finally, the above notation can be written as

$$E[e^2(n)] = P_d - 2R_{dX}^T H + H^T R_{XX} H$$

where

$$R_{dX} = E[d(n)X^T(n)]$$
$$R_{XX} = E[X(n)X^T(n)]$$
$$P_d = E[d^2(n)]$$

The vector R_{dX} is the cross-correlation between the desired signal $d(n)$ and the input signal $X(n)$. The R_{XX} matrix is the auto-correlation matrix of the input signal. To minimise $E[e^2(n)]$, the following equation should hold:

$$\frac{\partial E[e^2(n)]}{\partial h_i} = 0, \quad i = 0, 1, \ldots, N-1$$

Therefore

$$\begin{aligned}\frac{\partial E[e^2(n)]}{\partial h_i} &= \frac{\partial P_d}{\partial h_i} - 2\frac{\partial R_{dX}{}^T H}{\partial h_i} + \frac{\partial H^T R_{XX} H}{\partial h_i}, \qquad i = 0, 1, \ldots, N-1 \\ &= -2R_{dX} + 2R_{XX}H \\ &= 0\end{aligned}$$

Finally, the optimum coefficient vector is

$$H_{opt} = \frac{R_{dX}}{R_{XX}} = R_{dX}R_{XX}^{-1} \qquad [7.3]$$

Equation [7.3] is known as the Wiener–Hopf equation. It is evident that for real-time applications, the calculation of H_{opt} is time consuming and impractical as it involves a matrix inversion.

7.3 Least mean square (LMS)

The basic premise of the LMS algorithm is the use of the steepest descent algorithm shown below:

$$h_n(k) = h_{n-1}(k) + \beta\Delta_{n,k}$$

β is a positive value known as the step size parameter and $\Delta_{n,k}$ is a gradient vector that makes $H(n)$ approach the optimal value H_{opt}. It has been shown (Widrow and Stearns, 1985) that

$$\Delta_{n,k} = e(n)x(n-k)$$

Finally,

$$h_n(k) = h_{n-1}(k) + \beta e(n)x(n-k)$$

7.4 Implementation of an adaptive filter using the LMS algorithm

In this section, the adaptive filter shown in Figure 7.1 is considered by using an FIR digital filter and the LMS algorithm. Figure 7.2 shows the procedure for implementation. Note that the initialisation is done only once. The complete algorithm is shown in Program 7.1. The main program is similar

```
#include <stdio.h>
#include <string.h>
#include <common.h>
#include <codec.h>
#include <mcbspdrv.h>
#include <board.h>
#include <stdlib.h>
#include <timer.h>
#include <intr.h>
#include <regs.h>

#define SAMPLING_RATE 20000

extern int inicodec (int sampling_rate);

short h[8]={0,0,0,0,0,0,0,0};
short X[8]={0,0,0,0,0,0,0,0};
int Y=0;
int E=0;

#pragma CODE_SECTION (LMS_isr, ".iprog")
interrupt void LMS_isr (void)
{
  int N=8;
  int i, temp;
  short BETA_E,D;
  int beta = 0x00000174;

  temp = MCBSP0_DRR;

  X[0] = (short) temp;
  D = X[0];

  Y=0;

  for(i=0;i<N;i++)
    Y = Y + ((_mpy(h[i],X[i])) << 1) ;
    E = D -(short) (Y>>16);
    BETA_E =(short)((_mpy(beta,E)) >>15);

    for(i=N-1;i>=0;i--)
    {
       h[i] = h[i] +((_mpy(BETA_E,X[i])) >> 15);
       X[i]=X[i-1];
    }
```

Program 7.1 Program for the LMS algorithm

Program 7.1 continued

```
    MCBSP0_DXR = (temp &0xffff0000) |( ( (short)(Y>>16) )& 0x0000ffff);

    return;
}

void Init_Interrupts ()
{
  intr_reset();                     // Reset the interrupt system,
                                    // disable all interrupts.
  INTR_CLR_FLAG (CPU_INT15);        // Clear previous interrupt request
  INTR_ENABLE (CPU_INT15);          // Enable cpu interrupt line 15
  intr_map(CPU_INT15,ISN_RINT0);    // Link receive interrupt of serial
                                    // port 0 to interrupt line 15.
  intr_hook (LMS_isr,CPU_INT15);    // Assign the interrupt service routine
}

void main(void)
{
  evm_init();                       // Initialise EVM board.
  inicodec (SAMPLING_RATE);         // Initialise codec, adjust sampling rate.
  Init_Interrupts ();               // Initialise interrupts and hook isr.
  INTR_GLOBAL_ENABLE();             // Enable global interrupt.

  for (;;);                         // Main loop, does nothing.
}
```

to the one used for the FIR and IIR. The complete program can be found in directory `F:\DSPCODE\Adaptive (Chap7)\LMS_C\`.

This program can alternatively be written directly in assembly language, which enables the algorithm to be executed in fewer cycles. To further increase the speed of the algorithm, the coefficients and the delayed values of input values are all stored in registers, preventing the need for time-consuming memory access. However, if the algorithm is implemented in this manner the registers must not be used in between servicing the interrupt. If any are to be used, then they must be saved and restored every time the interrupt is serviced, otherwise the algorithm will not function correctly.

In the following example the assembly code is derived from the C subroutine shown in Program 7.1. Line by line, each C command is converted into assembly. The first lines to be converted are those that perform the FIR filtering, as shown below. Registers A1 to A8 are used to store the eight inputs to the filter, with the value in A1 being the most recent and A8 being the most delayed value. Registers B1 to B8 are used to store the filter coefficients with $h(0)$ stored in B1 and $h(7)$ in B8. The filtering function is then performed by simply multiplying B1 by A1, B2 by A2, etc., and adding the result of all of these products (see Figure 7.3).

The FIR C code can be written as:

```
for (i=0; i<8; i++)  y=y+h[i]x[i];
```

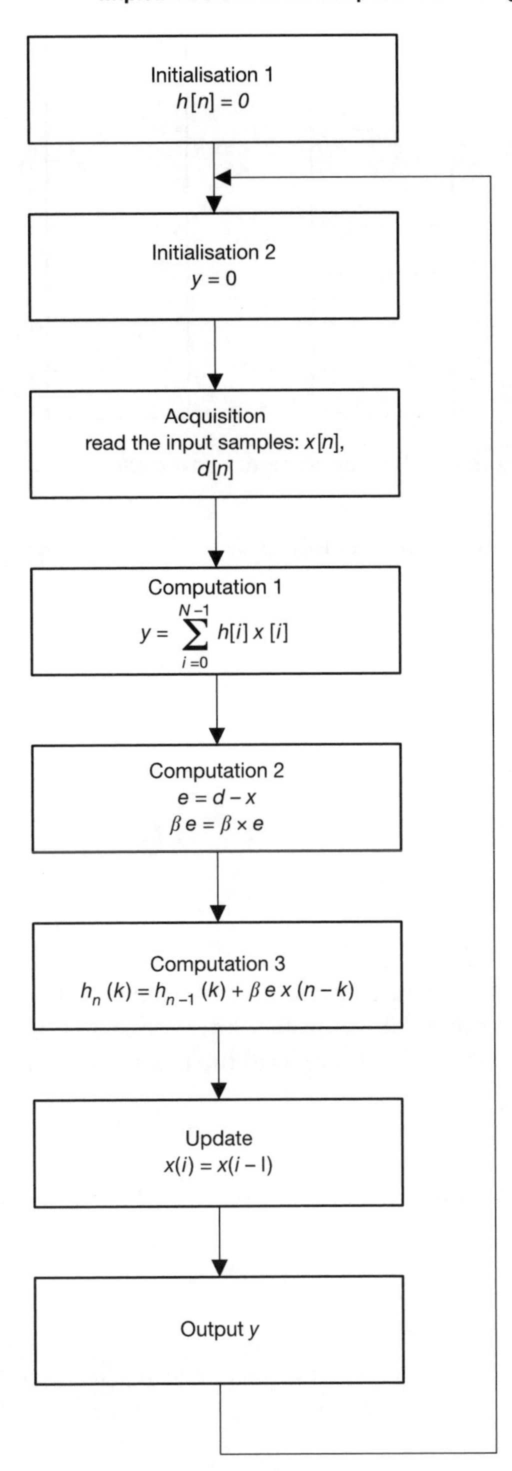

Figure 7.2 Steps for implementing an LMS adaptive filter

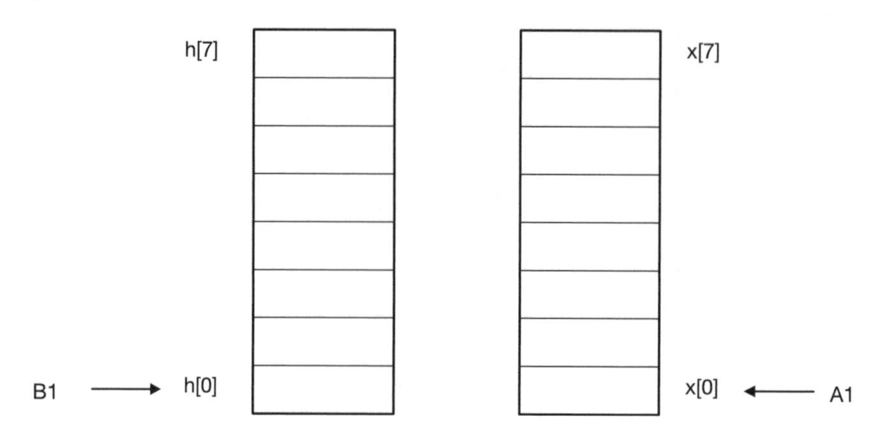

Figure 7.3 Registers used for coefficients and data

This can be translated into assembly as:

```
MPYH   B8,A8,A10
NOP
ADD    A10,A11,A11
MPYH   B7,A7,A10
NOP
ADD    A10,A11,A11
.
.
.
MPYH   A1,B1,A10
NOP
ADD    A10,A11,A11

SHL    A11,1,A11
```

The accumulated value is shifted in order to preserve the correct representation in fixed point format. However, the shift is done after all values have been summed (as shown above) as this reduces the number of shifts required to one, rather than eight.

The code shown above can be optimised further by replacing the `NOP`s with useful instructions. Note that to be able to do this, some registers have to be changed since instructions are put in the pipeline. The optimised code is shown below:

```
MPYH   B8,A8,A10
MPYH   B7,A7,B10
ADD    A10,A11,A11    ; A11 is used as an accumulator for the even registers
ADD    B10,B11,B11    ; B11 is used as an accumulator for the odd registers

MPYH   B6,A6,A10
MPYH   B5,A5,B10
ADD    A10,A11,A11
ADD    B10,B11,B11
```

```
MPYH    B4,A4,A10
MPYH    B3,A3,B10
ADD     A10,A11,A11
ADD     B10,B11,B11

MPYH    B2,A2,A10
MPYH    B1,A1,B10
ADD     A10,A11,A11
ADD     B10,B11,B11
ADD     A11,B11,A11
SHL     A11,1,A11
```

An extra ADD is now required at the end to combine A11 and B11 which each hold the summation of the output from the even and odd numbered taps respectively. This ADD is finally followed by the shift.

Since some multiplication and related ADD instructions are independent of each other, the previous code can be reordered as follows, with the multiplications and additions being in parallel.

```
        MPYH    B8,A8,A10
||      MPYH    B7,A7,B10

        MPYH    B6,A6,A10
||      MPYH    B5,A5,B10
        ADD     A10,A11,A11
||      ADD     B10,B11,B11

        MPYH    B4,A4,A10
||      MPYH    B3,A3,B10

        MPYH    B2,A2,A10
||      MPYH    B1,A1,B10
        ADD     A10,A11,A11
||      ADD     B10,B11,B11
        ADD     A10,A11,A11
||      ADD     B10,B11,B11

        ADD     A11,B11,A11

        SHL     A11,1,A11
```

The next line of C code to be converted is the calculation of the error.

```
E=D-(short)(y>>16);
  SUB   A0,A11,A11    ; The error is now stored in A11 (Note the desired value is in A0)
```

Now that the error has been calculated, the next task is to update the filter coefficients based on the error. However, before this is done, it is a good idea to calculate the product of β and the error outside the loop in order to save cycles within the loop. This line of C becomes the following in assembly:

```
BETA_E=(short)((_mpy(beta,E))>>15);
  MPYH  A11,A12,A11  ; Note that the value of β is loaded into A12 in the
                     ; initialisation routine
  SHL   A11,1,A11    ; A11 now holds the product of the error and β
```

Finally, the update of the filter coefficients can take place. As the C code shows, this involves multiplying `BETA_E` by each value of x and adding the shifted result to the coefficient that relates to that value of x.

Before the whole algorithm is repeated again, all the values of x need to be shifted so that the oldest value is discarded in order that a new one can be added. This is done by moving A7 to A8, A6 to A7, A5 to A6, ..., A1 to A2. However, this shift can only be done after the values of x have been used in the update equation. In the following section of code that performs the update, the required moves of the values of x are included in parallel with other instructions, after the value is no longer required.

The following code is written in a linear form so that it is easy to follow.

```
        MV      A11,B11     ; To allow the use of both paths

        MPYH    B11,A8,B10
||      MPYH    A11,A7,A10
        NOP
        SHL     B10,1,B10
||      SHL     A10,1,A10
        ADD     B10,B8,B8   ; h_n(7) = h_{n-1}(7) + (BETA_E * x(n - 7)) << 1
||      ADD     A10,B7,B7   ; h_n(6) = h_{n-1}(6) + (BETA_E * x(n - 6)) << 1
||      MV      A7,A8

        MPYH    B11,A6,B10
||      MPYH    A11,A5,A10
        NOP
        SHL     B10,1,B10   ; h_n(5) = h_{n-1}(5) + (BETA_E * x(n - 5)) << 1
||      SHL     A10,1,A10   ; h_n(4) = h_{n-1}(4) + (BETA_E * x(n - 4)) << 1
        ADD     B10,B6,B6
||      ADD     A10,B5,B5
||      MV      A6,A7

        MPYH    B11,A4,B10
||      MPYH    A11,A3,A10
||      MV      A5,A6
        NOP
        SHL     B10,1,B10
||      SHL     A10,1,A10
        ADD     B10,B4,B4   ; h_n(3) = h_{n-1}(3) + (BETA_E * x(n - 3)) << 1
||      ADD     A10,B3,B3   ; h_n(2) = h_{n-1}(2) + (BETA_E * x(n - 2)) << 1
||      MV      A4,A5

        MPYH    B11,A2,B10
||      MPYH    A11,A1,A10
||      MV      A3,A4
        NOP
        SHL     B10,1,B10
||      SHL     A10,1,A10
||      MV      A2,A3
        ADD     B10,B2,B2   ; h_n(1) = h_{n-1}(1) + (BETA_E * x(n - 1)) << 1
||      ADD     A10,B1,B1   ; h_n(0) = h_{n-1}(0) + (BETA_E * x(n)) << 1
||      MV      A2,A1
```

The next obvious step is to remove the NOPs and replace them with useful instructions. To do this, the use of another two registers is required, allowing two groups of parallel multiply instructions to be pipelined. This results in the following modified code.

```
        MPYH  B11,A8,B10
||      MPYH  A11,A7,A10

        MPYH  B11,A6,B9     ; use of B9 instead of B10
||      MPYH  A11,A5,A9     ; use of A9 instead of A10

        SHL   B10,1,B10
||      SHL   A10,1,A10

        ADD   B10,B8,B8
||      ADD   B7,A10,B7
||      MV    A7,A8

        SHL   B9,1,B9
||      SHL   A9,1,A9

        ADD   B9,B6,B6
||      ADD   A9,B5,B5
||      MV    A6,A7

        MPYH  B11,A4,B10
||      MPYH  A11,A3,A10
||      MV    A5,A6

        MPYH  B11,A2,B9
||      MPYH  A11,A1,A9
||      MV    A4,A5

        SHL   B10,1,B10
||      SHL   A10,1,A10

        ADD   B10,B4,B4
||      ADD   A10,B3,B3
||      MV    A3,A4

        SHL   B9,1,B9
||      SHL   A9,1,A9
||      MV    A2,A3

        ADD   B9,B2,B2
||      ADD   A9,B1,B1
||      MV    A1,A2
```

All the optimised sections shown can now be combined with the initialisation section and the command required to load the new value of x and the corresponding desired value d from the codec. The complete assembly listing follows (see Program 7.2).

Note that the upper 16 bits of the value read from the codec represent the signal x and the lower 16 bits represent the desired signal d. Also note that a section is included to write the desired value and the output from the filter to the codec allowing observation of the performance of the filter in terms of how the desired and the actual responses compare with time. The complete program can be found in directory `F:\DSPCODE\Adaptive (Chap7)\LMS_ASM\`.

```
        .def    _lms_asm_init
        .def    _lms_asm_intr

        .sect   "iProg"

        .ref    _h
        .ref    _x_in
        .ref    _BETA

_lms_asm_init
        MVKL    0x018C0000,B12  ; Store Address of CODEC in B12
        MVKH    0x018C0000,B12

        MVKL    _BETA, A12      ; Store Address of BETA in A12
        MVKH    _BETA, A12
        ZERO    B1              ; Zero ALL Filter Coefficients
        ZERO    B2
        ZERO    B3
        ZERO    B4
        ZERO    B5
        ZERO    B6
        ZERO    B7
        ZERO    B8

        LDH     *A12, A12       ; Load the value of BETA into A12

        MVKL    0x0000ffff,A15  ; Load A15 with a 16 LSB mask
        MVKH    0x0000ffff,A15

        MVKL    0xffff0000,A14  ; Load A14 with a 16 MSB mask
        MVKH    0xffff0000,A14

        MVC     CSR,B0
||      SHL     A12, 16, A12    ; Shift BETA into upper 16 bits
        OR      0X01,B0,B0      ; Set bit in B0 that relates to the GIE bit in the CSR
        MVC     B0,CSR          ; By writing this value the interrupts are enabled

loop    B       loop            ; Wait for an interrupt
        nop     5

_lms_asm_intr
        zero    A11
        zero    B11

        LDW     *B12,A1         ; Read the first element: 16 MSB = x[n], 16 LSB = d[n]
        nop     5

        SHL     A1, 16, A0      ; A0 contains the desired value in 16 MSB

        MPYH    B8,A8,A10
||      MPYH    B7,A7,B10

        MPYH    B6,A6,A10
||      MPYH    B5,A5,B10
```

Program 7.2 Assembly code for LMS algorithm and the required initialisation

Program 7.2 continued

```
        ADD     A10,A11,A11
||      ADD     B10,B11,B11

        ADD     A10,A11,A11
||      ADD     B10,B11,B11

        MPYH    B4,A4,A10
||      MPYH    B3,A3,B10

        MPYH    B2,A2,A10
||      MPYH    B1,A1,B10

        ADD     A10,A11,A11
||      ADD     B10,B11,B11

        ADD     A10,A11,A11
||      ADD     B10,B11,B11

        ADD     A11,B11,A11
        SHL     A11,1,A11
        MV      A11,B11

*Prepare the data to send out
        SHR     B11, 16, B11    ; Output from Filter is now in 16 LSB in B11
        AND     B11,A15,B11     ; Mask B11
        AND     A0,A14,A10      ; Mask A0 which holds the Desired Value
        ADD     A10,B11,A10     ; Combines the Output in the lower 16 bits and the desired
                                ; value in the upper 16 bits
        STW     A10,*B12[1]     ; Send the value in A10 to the CODEC

* E = D-SHORT(Y)>>15
        SUB     A0,A11,A11

* _mpy(beta,E)>>15
        MPYH    A11,A12,A11
        NOP
        SHL     A11, 1, A11
        MV      A11,B11

*h[i] = h[i] + ((_mpy(BETA_E, X[i])) >>15
        MPYH    B11,A8,B10
||      MPYH    A11,A7,A10

        MPYH    B11,A6,B9
||      MPYH    A11,A5,A9

        SHL     B10,1,B10
||      SHL     A10,1,A10

        ADD     B10,B8,B8
||      ADD     B7,A10,B7
||      MV      A7,A8

        SHL     B9,1,B9
||      SHL     A9,1,A9

        ADD     B9,B6,B6
||      ADD     A9,B5,B5
||      MV      A6,A7
```

Program 7.2 continued

```
        MPYH  B11,A4,B10
||      MPYH  A11,A3,A10
||      MV    A5,A6

        MPYH  B11,A2,B9
||      MPYH  A11,A1,A9
||      MV    A4,A5

        SHL   B10,1,B10
||      SHL   A10,1,A10

        ADD   B10,B4,B4
||      ADD   A10,B3,B3
||      MV    A3,A4

        SHL   B9,1,B9
||      SHL   A9,1,A9
||      MV    A2,A3

        ADD   B9,B2,B2
||      ADD   A9,B1,B1
||      MV    A1,A2

        b     .s2    IRP
        nop   5
```

It is possible to optimise this code further by placing more instructions in parallel and changing a few registers around. However, this makes the structure of the code more complicated and hence the implementation harder to follow; for this reason it is not included, although it is not hard to optimise further with a little careful thought. Figure 7.4 shows the set-up for testing the implemented adaptive filter and Figure 7.5 shows the converging output of the adaptive filter. To observe the converging output on a storage scope or a spectrum analyser, the following steps should be followed, otherwise the initialisation of the serial port will trigger the scope and the converging signal will not be captured.

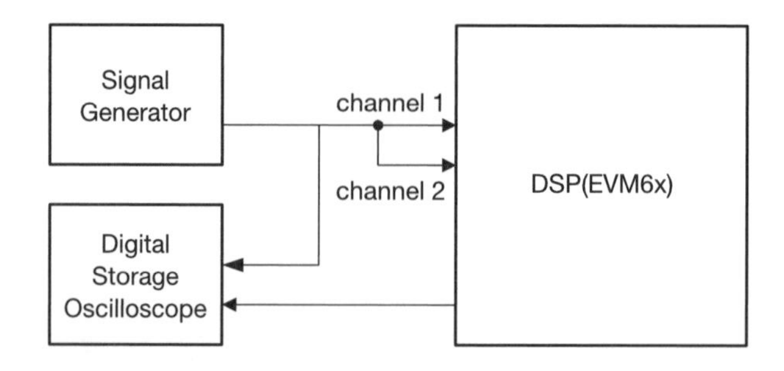

Figure 7.4 Test configuration for the adaptive filter

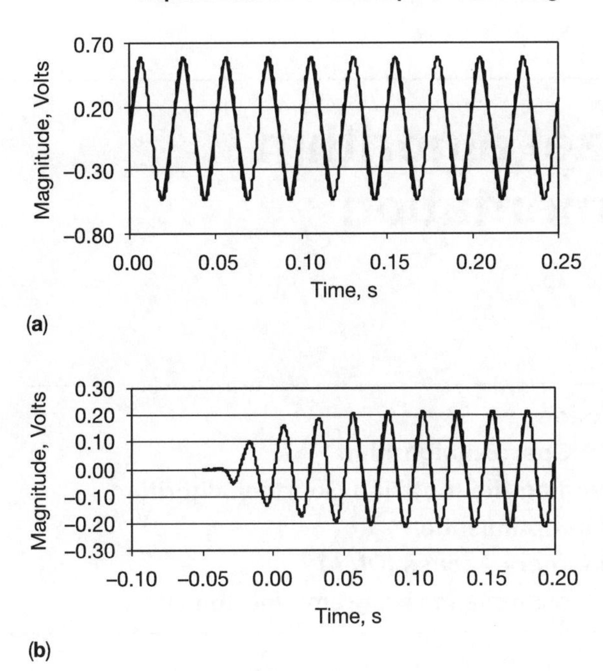

Figure 7.5 (a) Input and (b) converging output of the implemented adaptive filter

Using the debugger:

(1) Load the code to the EVM.
(2) Enter '`go main`' in the command line.
(3) Enter '`go lms_asm_intr`' in the command line.
(4) Set the storage scope or the spectrum analyser to single shot.
(5) Run the code.

Chapter 8

Goertzel algorithm implementation

8.1 Introduction

The Goertzel algorithm is mainly used to detect tones for Dual Tone Multi-Frequency (DTMF) applications. DTMF signalling is predominantly used for push-button digital telephone sets which are an alternative to rotary telephone sets, which are cumbersome and slow to use. The application of DTMF has now been extended to electronic mail and telephone banking systems in which users can select options from a menu by selecting DTMF signals from the telephone.

In a DTMF signalling system, a combination of two frequency tones represents a specific digit, a character or symbol (* or #). Two types of signal processing are involved with DTMF signals. These are coding (or generation) and decoding (or detection). For the coding, two sinusoidal sequences of finite length are added together to represent a digit, a character or a symbol as shown in Figure 8.1. For example, if the button '5' is pressed the 770 Hz and 1336 Hz tones will be generated.

DTMF detection involves the extraction of tones present in the signal followed by determination of the corresponding digit, character or symbol. Decoding DTMF signals can be accomplished either by hardware or by software techniques. Since the DTMF coding or decoding is normally part of a system, implementation in software reduces the chip count and cost, and provides better stability and precision.

This chapter deals only with DTMF detection and provides a practical example of the Goertzel algorithm. This chapter also shows how to produce

	Column 1 1209Hz	Column 2 1336Hz	Column 3 1477Hz	Column 4 1633Hz
Row 1 697 Hz	1	2	3	A
Row 2 770 Hz	4	5	6	B
Row 3 852 Hz	7	8	9	C
Row 4 941 Hz	*	0	#	D

Figure 8.1 Frequency distribution for a standard telephone keypad

optimised code by the pen and paper method and by using linear assembly, and also demonstrates how to program the Direct Memory Access (DMA).

8.2 Modified Goertzel algorithm

The Goertzel algorithm is derived from the Discrete Fourier Transform (DFT) and exploits the periodicity of the phase factor $\exp(-j.2\pi k/N)$ to reduce the computational complexity associated with the DFT, as the Fast Fourier Transform (FFT) does. However, the Goertzel algorithm is more efficient than the FFT, when only a few samples of the DFT are required. With the Goertzel algorithm only 16 samples of the DFT are required. The Goertzel algorithm can be derived as follows using the DFT equation.

$$X(k) = \sum_{n=0}^{N-1} x(n) e^{-j\frac{2\pi kn}{N}}$$

$$= e^{j\frac{2\pi kN}{N}} \sum_{n=0}^{N-1} x(n) e^{-j\frac{2\pi kn}{N}}, \quad e^{j\frac{2\pi kN}{N}} = 1 \qquad [8.1]$$

$X(k)$ can be written as

$$X(k) = \sum_{n=0}^{N-1} x(n) h_k(N-n)$$

where

$$h_k(N-n) = e^{-j\frac{2\pi k}{N}(n-N)}$$

$X(k)$ can now be identified as an output of a filter at $m = N$ and the filter can be written as follows:

$$y_k(m) = \sum_{n=0}^{N-1} x(n)h_k(m-n) \qquad [8.2]$$

Therefore:

$$X_k(N) = y_k(m)|_{m=N} \qquad [8.3]$$

This can be read as: the Discrete Fourier Transform $X_k(N)$ is the output of the filter $y_k(m)$ at $m = N$.

The z-transform of y_k, $y_k(z)$, is given by

$$y_k(z) = \frac{X(z)}{1 - e^{j\frac{2\pi k}{N}} z^{-1}} \qquad [8.4]$$

Deriving the difference equation,

$$y_k(n) = x(n) + e^{j\frac{2\pi k}{N}} y_k(n-1), \quad y_k(-1) = 0 \qquad [8.5]$$

This can be implemented as a single-pole IIR filter as shown in Figure 8.2.

Equation [8.5] involves multiplication by a complex number and each complex multiplication results in four real multiplications and four real additions.

To avoid complex multiplication, Equation [8.4] is multiplied by a complex conjugate pole and simplified as follows:

$$\begin{aligned} y_k(z) &= \frac{X(z)\left(1 - e^{-j\frac{2\pi}{N}k} z^{-1}\right)}{\left(1 - e^{j\frac{2\pi}{N}k} z^{-1}\right)\left[\left(1 - e^{-j\frac{2\pi}{N}k} z^{-1}\right)\right]} \\ &= X(z)\frac{\left(1 - e^{-j\frac{2\pi}{N}k} z^{-1}\right)}{1 - 2\cos\left(\frac{2\pi}{N}k\right) z^{-1} + z^{-2}} \end{aligned} \qquad [8.6]$$

Equation [8.6] can be implemented using the structure shown in Figure 8.3 (see Chapter 6), where

$$Q_n = x(n) + 2\cos\left(\frac{2\pi k}{N}\right) Q_{n-1} - Q_{n-2} \qquad [8.7]$$

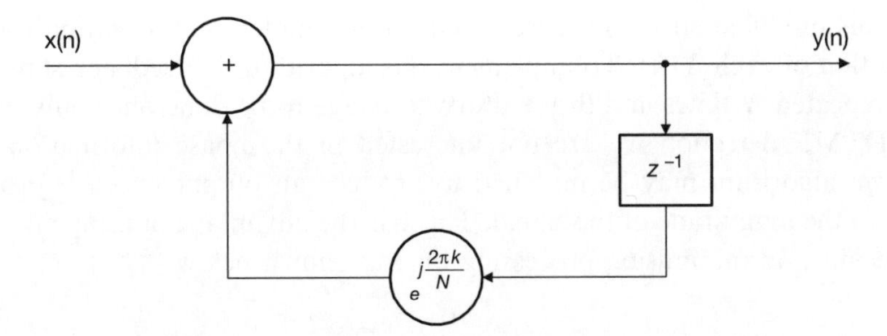

Figure 8.2 Structure of a single-pole resonator for computing the DFT

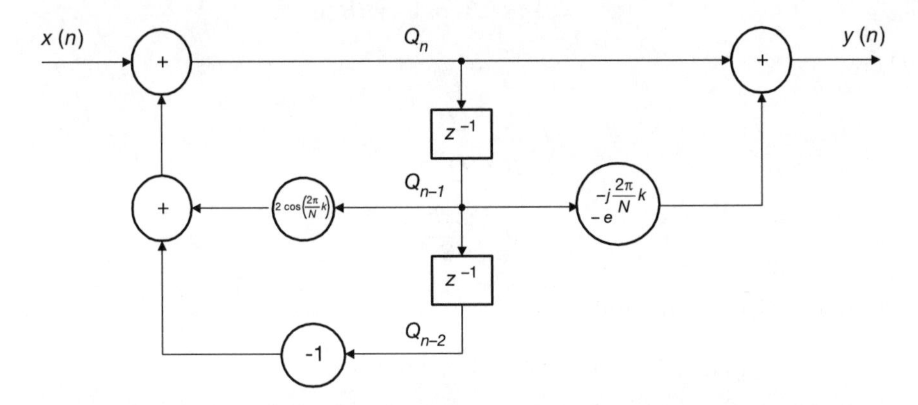

Figure 8.3 Structure of a two-pole resonator for computing the *k*th DFT sample

and

$$y_k(n) = Q_n - Q_{n-1}e^{-j\frac{2\pi}{N}k} \tag{8.8}$$

or

$$X_k(N) = y_k(m)\big|_{m=N}$$

This means that $y_k(n)$ need not be calculated for every value of m as it only needs to be calculated at $m = N$.

Finally Equations [8.7] and [8.8] can be rewritten as follows:

$$Q_n = x(n) + 2\cos\left(\frac{2\pi k}{N}\right)Q_{n-1} - Q_{n-2}, \quad 0\leqslant n\leqslant N \tag{8.9}$$

and

$$y_k(n)\big|_{n=N} = X(k) = Q_N - Q_{N-1}e^{-j\frac{2\pi}{N}k}, \quad n = N \tag{8.10}$$

With this modified structure there is only one complex multiplication for the calculation of each $X(k)$. To implement this algorithm the feedback stage has to be repeated N times and the feedforward stage to be done once only. Since with DTMF detection we are not interested in the phase information, the Goertzel algorithm may be modified to produce an output which is proportional to the magnitude of the signal. This has the advantage of using only real coefficients and minimising processing time as shown below.

$$\begin{aligned} y_k(N) &= Q_N - e^{-j\frac{2\pi}{N}k} Q_{N-1} \\ &= A - Be^{-j\theta} \\ &= [A - B\cos\theta] + jB\sin\theta \end{aligned}$$

where

$$\begin{aligned} A &= Q_n \\ B &= Q_{n-1} \\ \theta &= \frac{2\pi k}{N} \end{aligned}$$

The square of the magnitude is

$$\begin{aligned} |y_k(n)|^2 &= (A - B\cos\theta)^2 + (B\sin\theta)^2 \\ &= A^2 - 2AB\cos\theta + B^2\cos^2\theta + B^2\sin^2\theta \\ &= A^2 + B^2 - 2AB\cos\theta \end{aligned}$$

Therefore:

$$\begin{aligned} |y_k(N)|^2 &= |X(k)|^2 \\ &= Q^2(N) + Q^2(N-1) - 2\cos\left(\frac{2\pi k}{N}\right) Q(N)Q(N-1) \qquad [8.11] \end{aligned}$$

Equation [8.11] shows that the calculation of $|X(k)|^2$ requires only real additions and multiplications.

8.3 Implementing the modified Goertzel algorithm

To implement the modified Goertzel algorithm, the following are required:

(1) Equations [8.9] and [8.11]
(2) The filter structure

(3) Values of k and N
(4) Sampling rate and frequency to be detected.

Equations [8.9], [8.11], and the filter structure have already been derived in the previous section. k is an integer corresponding to the frequency, f_0, to be detected. k and f_0 are related by the following equations:

$$\theta_k = \frac{2\pi}{N}k \qquad \text{for integer } k$$

or

$$\theta_k = \frac{2\pi}{f_s}f_0$$

θ_k is the digital frequency of the analogue frequency f_0 to be detected, and f_s is the sampling frequency. Hence,

$$k = N\frac{f_0}{f_s}$$

The sampling frequency is set to 8000 Hz by the telephone system, and the tones are predefined by Bell Laboratories as the DTMF tones. The only parameter left is N, and this is set to be 205 for the eight DTMF tones. Table 8.1 shows the relationship between the tones to be detected, the value of k and the coefficient $2\cos(2\pi k/N)$ in both floating point and Q15 format.

Program 8.1 shows the implementation of the modified Goertzel algorithm. In this program the tone to be detected is 1209 Hz. Note that some coefficients are greater than 1 and are therefore divided by 2 before being converted to Q15 format. This is taken into account in the program by shifting the product of 'delay_1 *coef_1' by 14 and not 15. The input signal is scaled down to prevent overflow occurring. Since the modified Goertzel algorithm produces the magnitude squared of the signal every 205 samples, it is not possible to display it on the oscilloscope because of the DC blocking capacitor at the

Fundamental frequency (Hz)	k	Coefficient (decimal)	Coefficient (Q15)
697	18	1.703275	6D02*
770	20	1.635585	68AD*
852	22	1.562297	63FC*
941	24	1.482867	5EE7*
1209	31	1.163138	4A70*
1336	34	1.008835	4090*
1477	38	0.790074	6521
1633	42	0.559454	479C

* The decimal values are divided by 2 in order to be represented in Q15 format. This has to be taken into account during implementation.

Table 8.1 Coefficients in decimal and Q15 formats ($N = 205$, $f_s = 8000$ Hz)

output of the codec (see Chapter 3). To get around this problem the output of the modified Goertzel algorithm is modulated by the input signal before being sent to the codec. Figure 8.4 shows the set-up for testing the algorithm. The frequency response of the algorithm is shown in Figure 8.5. The complete program can be found in directory `F:\DSPCODE\Goertzel (Chap8)\BASIC\`.

```
/*-----------------------------------------------------------------------
  File:         gtz.c
  Description: Perform goertzel algorithm on 200 samples of codec input.
               Sends goetzel value periodically to mailbox.
               Each sample generates an interrupt;
               after each 200 interrupts, a goertzel value is generated.
-----------------------------------------------------------------------*/
#include <stdio.h>
#include <string.h>
#include <common.h>
#include <codec.h>
#include <mcbspdrv.h>
#include <board.h>
#include <stdlib.h>
#include <timer.h>
#include <intr.h>
#include <regs.h>

#define SAMPLING_RATE 8000

extern int inicodec (int sampling_rate);

#pragma CODE_SECTION (isr_rint0, ".iprog")
interrupt void isr_rint0 (void)
{
  static short delay;
  static short delay_1 = 0;
  static short delay_2 = 0;
  static int N = 0;
  static int Goertzel_Value = 0;

  int i;
  int prodl, prod2, prod3, sum;
  int R_in;
  short input;
  int output;
  short coef_1 = 0x4A70;       // For detecting 1209 Hz

  R_in = MCBSP0_DRR;

  input = (short) R_in;
  input = input >> 4;    // Scale down input to prevent overflow.

  prodl = (delay_1*coef_1)>>14;
  delay = input + (short)prodl - delay_2;
  delay_2 = delay_1;
  delay_1 = delay;
  N++;
```

Program 8.1 C code implementation of the Goertzel algorithm

Program 8.1 continued

```
  if (N==206)
  {
    prod1 = (delay_1 * delay_1);
    prod2 = (delay_2 * delay_2);
    prod3 = (delay_1 *  coef_1)>>14;
    prod3 = prod3 * delay_2;
    Goertzel_Value = (prod1 + prod2 - prod3) >> 15;
    Goertzel_Value <<= 4;          // Scale up value for sensitivity.
    N = 0;
    delay_1 = delay_2 = 0;
  }

  output = (((short) R_in) * ((short)Goertzel_Value)) >> 15;
  MCBSP0_DXR = (output<<16) | (output & 0x0000ffff);

  return;
}

void Init_Interrupts ()
{
  intr_reset();                     // Reset the interrupt system,
                                    // disable all interrupts.
  INTR_CLR_FLAG (CPU_INT15);        // Clear previous interrupt request
  INTR_ENABLE (CPU_INT15);          // Enable cpu interrupt line 15
  intr_map(CPU_INT15,ISN_RINT0);    // Link receive interrupt of serial
                                    // port 0 to interrupt line 15.
  intr_hook (isr_rint0,CPU_INT15); // Assign the interrupt service routine
}

void main(void)
{
  evm_init();                       // Initialise EVM board.
  inicodec (SAMPLING_RATE);         // Initialise codec, adjust sampling rate.
  Init_Interrupts ();               // Initialise interrupts and hook isr.

  INTR_GLOBAL_ENABLE();             // Enable global interrupt.

  for (;;);                         // Main loop, does nothing.
}
```

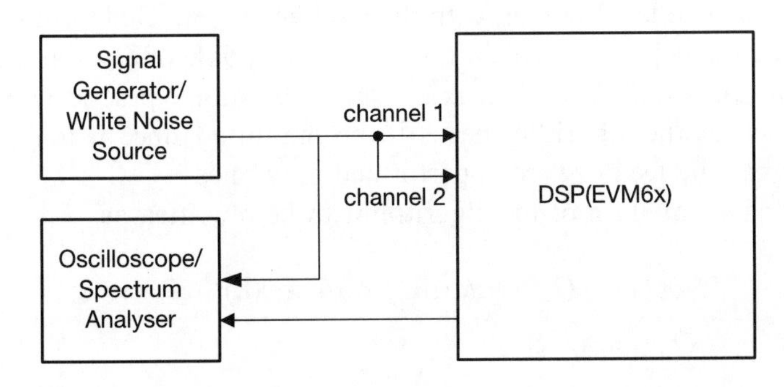

Figure 8.4 Set-up for testing the modified Goertzel algorithm

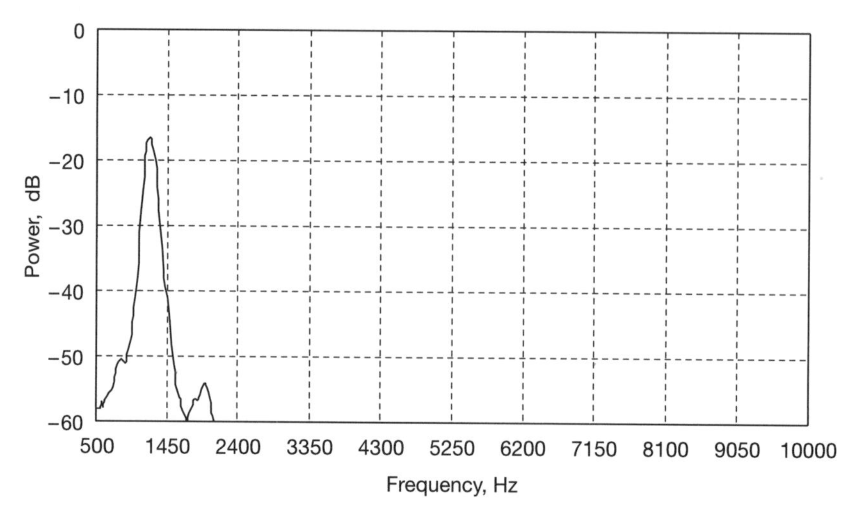

Figure 8.5 Frequency response of the Goertzel algorithm for a 1209 Hz tone

8.4 Algorithm optimisation

The previous code (Program 8.1) has been benchmarked and it takes 52 cycles to perform one iteration for the calculation of Q_n, Q_{n-1} and Q_{n-2} (see Figure 8.6). In order to optimise this code we need to change the operation of the algorithm to introduce loops (see Chapter 4). To do this we need to first get the N samples and then perform the feedback and feedforward operations.

To further accelerate the process, we can use the DMA to collect the data samples and therefore free the CPU. First let us concentrate on optimising the Goertzel algorithm.

8.4.1 Hand optimisation of the Goertzel algorithm

It has been shown in Chapter 4 that using the dependency diagram can help in hand-optimising code. However, with the modified Goertzel algorithm it is not simple to use the dependency diagram due to the update of Q_{n-1} and Q_{n-2}. In this section, an intuitive method is shown to be more 'practical'. Only the feedback part of the algorithm needs to be optimised since it is repeated N times, whereas the feedforward is performed only once.

The feedback equation of the algorithm can be rewritten as

$$Q_n = (Q_{n-1} \times \text{coeff}) \gg 14 + x(n) - Q_{n-1} \qquad [8.12]$$

$$Q_{n-2} = Q_{n-1} \qquad [8.13]$$

$$Q_{n-1} = Q_n \qquad [8.14]$$

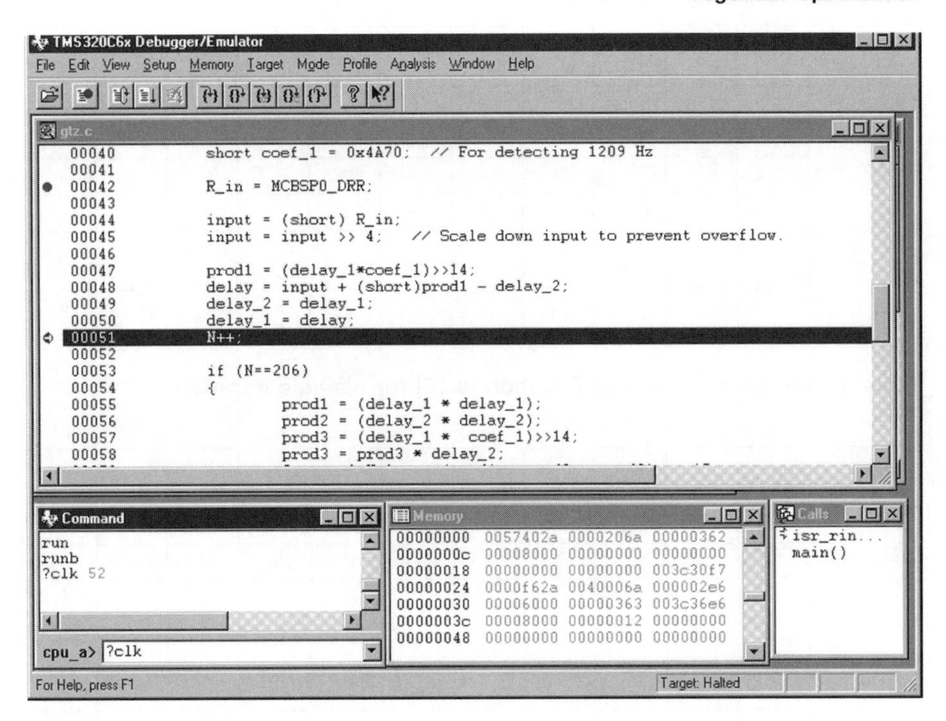

Figure 8.6 Screen dump showing the set-up used for benchmarking

Equation [8.14] needs to be evaluated when Equation [8.12] has been evaluated, whereas Equation [8.13] can be performed as soon as $(Q_{n-1} \times \text{coeff}) \gg 14$ is complete. Equation [8.12] can be evaluated as

$$Q_n = [(Q_{n-1} \times \text{coeff}) \gg 14 + u(n)] - Q_{n-2} \qquad [8.15]$$

or

$$Q_n = [(Q_{n-1} \times \text{coeff}) \gg 14] + [x(n) - Q_{n-2}] \qquad [8.16]$$

Although mathematically Equations [8.15] and [8.16] are similar when it comes to implementation, they lead to two different optimisations.

Tables 8.2 and 8.3 show one iteration of Equations [8.15] and [8.16] on a cycle-by-cycle basis. Both tables include Equations [8.13] and [8.14]. Notice that the data move $Q_{n-1} = Q_n$ is performed by the `ADD` instruction, saving one instruction. Also notice that using Equation [8.16] instead of Equation [8.15] reduces the number of cycles between the `LD` and `ADD`.

If we use Equation [8.16] and refer to Table 8.3, we can observe that another multiply can start just after the `ADD` instruction, in which case we need to load x two cycles before the `MPY`, etc. This is shown in Table 8.4 and Program 8.2.

Cycle/ iteration	1	2	3	4	5	6	7	8	9	10	11
1	LD										
						ADD					
							SUB				
			MPY		SHR						
							MV ($Q_{n-2} = Q_{n-1}$)				
								MV ($Q_{n-1} = Q_n$)			

Table 8.2 Representation of Equation [8.15] for a single iteration

Cycle/ iteration	1	2	3	4	5	6	7	8	9	10	11
1	LD										
						SUB					
							ADD ($Q_{n-1} = Q_n$)				
				MPY							
						SHR					
					MV ($Q_{n-2} = Q_{n-1}$)						

Table 8.3 Representation of Equation [8.16] for a single iteration

Cycle/ iteration	1	2	3	4	5	6	7	8	9	10	11
1	LDH										
						SUB					
							ADD				
				MPY							
						SHL					
						MV					
2										SUB	
											ADD
					LDH			MPY			
										SHL	
										MV	
3											
									LDH		

Table 8.4 Representation of Equation [8.16] for multi-iteration

```
; PIPED LOOP PROLOG

        LDH     .D1T1     *A0++(4),A3
        NOP
        NOP     1

        MPY     .M2       B4,B5,B6

        LDH     .D1T1     *A0++(4),A3

        MV      .L1X      B4,A4
||      SUB     .D1       A3,A4,A3
||      SHR     .S2       B6,0xe,B4

        ADD     .L2X      A3,B4,B4

        MPY     .M2       B4,B5,B6

        LDH     .D1T1     *A0++(4),A3

        MV      .L1X      B4,A4
||      SUB     .D1       A3,A4,A3
||      SHR     .S2       B6,0xe,B4

        ADD     .L2X      A3,B4,B4
         -
         -
         -
        MPY     .M2       B4,B5,B6

        LDH     .D1T1     *A0++(4),A3

        MV      .L1X      B4,A4
||      SUB     .D1       A3,A4,A3
||      SHR     .S2       B6,0xe,B4

        ADD     .L2X      A3,B4,B4
```

Program 8.2 Assembly code implementation of Equation [8.16]

The code shown in Program 8.2 is unrolled. To roll it we need to use the branch, `B`, and `SUB` instructions and the problem becomes where to insert them. First we put a label (e.g. `loop`) in the location we wish to branch to. Next we insert the branch instruction into the kernel so that after five cycles it branches to the location of `loop`. The `SUB` instruction needs then to be placed just before the branch instruction. This is shown in Program 8.3.

```
; PIPED LOOP PROLOG

        LDH     .D1T1     *A0++(4),A3
        NOP
        NOP     1
```

Program 8.3 Intermediate assembly code for the implementation of Equation [8.16]

Program 8.3 continued

```
; PIPED  LOOP KERNEL

loop:    MPY   .M2     B4,B5,B6

   [ A1] SUB   .L1     A1,0x1,A1
||       LDH   .D1T1   *A0++(4),A3

         MV    .L1X    B4,A4
||       SUB   .D1     A3,A4,A3
||       SHR   .S2     B6,0xe,B4
|| [ A1] B     .S1     loop

         ADD   .L2X    A3,B4,B4
```

By examining Program 8.3 it is clear that the first time the `ADD` instruction is reached, the program does not branch to `loop`. Introducing a branch five cycles before the `ADD` and inserting a `SUB` instruction before the branch, as shown in Program 8.4, solves this problem.

```
; PIPED LOOP PROLOG

         LDH    .D1T1   *A0++(4),A3
|| [ A1] SUB    .L1     A1,0x1,A1

   [ A1] B      .S1     loop
         NOP    1

; PIPED LOOP KERNEL

loop:    MPY    .M2     B4,B5,B6

   [ A1] SUB    .L1     A1,0x1,A1
||       LDH    .D1T1   *A0++(4),A3

         MV     .L1X    B4,A4
||       SUB    .D1     A3,A4,A3
||       SHR    .S2     B6,0xe,B4
|| [ A1] B      .S1     loop

         ADD    .L2X    A3,B4,B4
```

Program 8.4 Final assembly code for the implementation of the Goertzel main loop

This shows that the feedback loop kernel can be optimised to four cycles.

8.4.2 Optimisation by using linear assembly

Linear assembly is a very powerful tool and can produce very good results if one has knowledge of the architecture of the processor and assembly language.

Program 8.5 shows how linear assembly can be written. The result obtained by linear assembly is comparable to that obtained by hand optimisation (see Program 8.6). The complete program can be found in directory F:\DSPCODE\Goertzel (Chap8)\GTZSA\.

```
          .def      _gz

          .sect"    .iprog"

_gz       .cproc    input, coeff, count

          .reg      delay1, delay2, x, gzv
          .reg      prod1, prod2, prod3, sum1, sum2

          zero      delay1
          zero      delay2

loop:
          ldw       *input++, x
          shr       x, 24, x

          mpy       delay1, coeff, prod1
          shr       prod1, 14, prod1

          sub       x, delay2, sum1
          mv        delay1, delay2

          add       sum1, prod1, delay1

[count]   sub       count,1,count
[count]   b         loop

          mpy       delay1, delay1, prod1
          mpy       delay2, delay2, prod2
          add       prod1, prod2, sum1

          mpy       delay1, coeff, prod3
          shr       prod3, 14, prod3
          mpy       prod3, delay2, prod3

          sub       sum1,prod3, sum1
          shr       sum1, 15, gzv

          .return   gzv

          .endproc
```

Program 8.5 Linear assembly code for the implementation of the Goertzel main loop

```
;** ----------------------------------------------------------------------*
L5:        ; PIPED LOOP PROLOG
           LDW    .D1T1    *A4++,A3      ; |38|
   [ B0]   SUB    .L2      B0,0x1,B0     ; |49|
   [ B0]   B      .S1      loop          ; |50|
           NOP             1
;** ----------------------------------------------------------------------*
```

Program 8.6 Output code generated by the linear assembly

Program 8.6 continued

```
loop:       ; PIPED LOOP KERNEL

            MPY    .M2     B5,B4,B6       ; ^ |41|
||          LDW    .D1T1   *A4++,A3       ; @ |38|

            SHR    .S1     A3,0x18,A3     ; |39|
|| [ B0]    SUB    .L2     B0,0x1,B0      ; @ |49|

            MV     .L1X    B5,A0          ; |45|
||          SUB    .D1     A3,A0,A3       ; |44|
||          SHR    .S2     B6,0xe,B5      ; ^ |42|
|| [ B0]    B      .S1     loop           ; @ |50|

            ADD    .L2X    A3,B5,B5       ; ^ |47|
;** -----------------------------------------------------------------*
L7:         ; PIPED LOOP EPILOG
            MPY    .M2     B5,B4,B6       ; @ ^ |41|
            SHR    .S1     A3,0x18,A3     ; @ |39|

            MV     .L1X    B5,A0          ; @ |45|
||          SUB    .D1     A3,A0,A3       ; @ |44|
||          SHR    .S2     B6,0xe,B5      ; @ ^ |42|

            ADD    .L2X    A3,B5,B5       ; @ ^ |47|
```

8.5 Direct Memory Access (DMA)

There are two means for transferring data from one part of the memory to another. The first is to use the CPU with the load and store instructions. The second is to use the DMA (see Figure 8.7). The DMA method has the advantage of transferring a block of data without using the CPU, therefore leaving it free to perform other tasks. Once the CPU core has specified data transfer options, it returns to normal operation and the DMA controller can operate as a stand-alone device for data transfer.

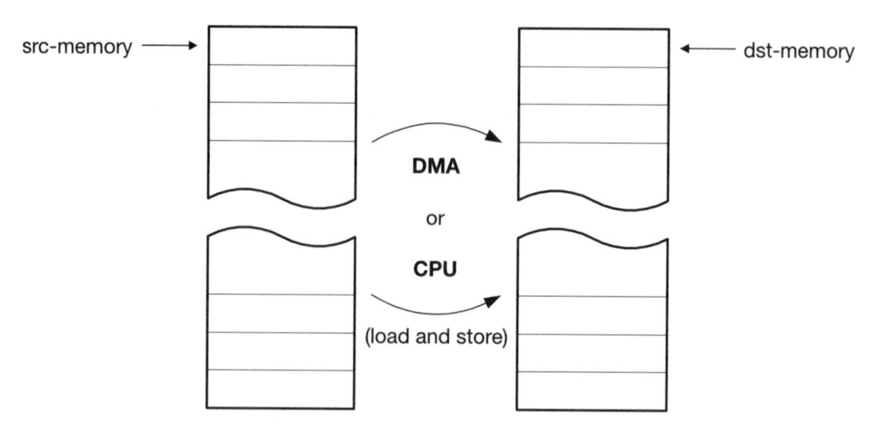

Figure 8.7 Possible means of data transfer

In this chapter a brief overview of the DMA operation is given, followed by a practical example. For more details on the DMA operation the reader may refer to the TMS320C6201/C6701 *Peripherals Reference Guide* (SPRU190).

8.5.1 DMA features

There are four DMA channels associated with the TMS320C6x processors plus one auxiliary channel for servicing requests from the Host Port Interface (HPI).

8.5.1.1 DMA channel priorities

To avoid memory conflicts when more than one DMA channel tries to access the same resource in a given cycle, a priority scheme is put into place. The four DMA channels have fixed priority, with channel 0 having the highest priority and channel 3 the lowest priority. However, the auxiliary channel may take any level of priority as shown in Figure 8.8. The priority is set by the DMA auxiliary control register shown in Figure 8.9. The one-bit AUXPRI field sets the priority between the DMA and the CPU, whereas the four-bit CH PRI field sets the priority of the auxiliary channel.

8.5.1.2 DMA operation and associated registers

For each DMA channel, there are five memory-mapped registers to configure. There are also five global memory-mapped registers which are common to all channels (see Figure 8.10).

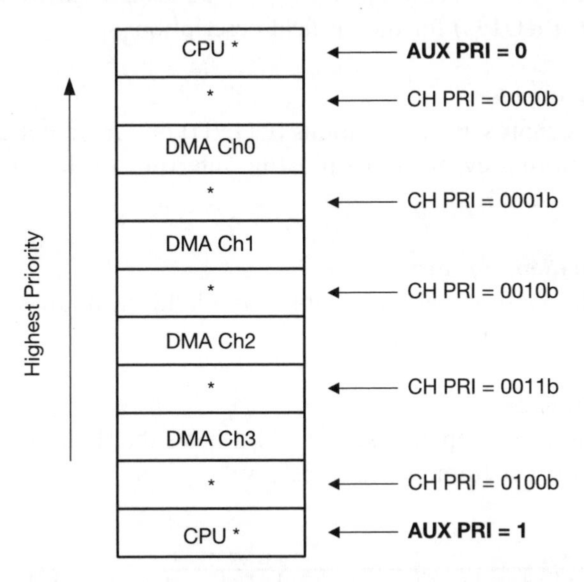

* : Possible priority of the auxiliary channel
CPU * : The CPU can have the first or the last priority

Figure 8.8 CPU–DMA priority

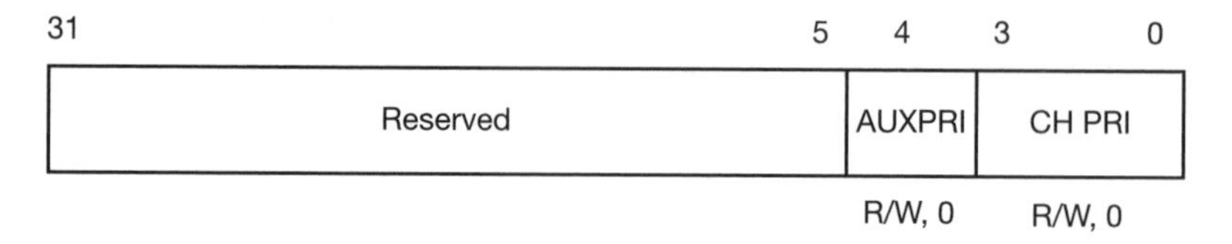

Figure 8.9 DMA auxiliary control register

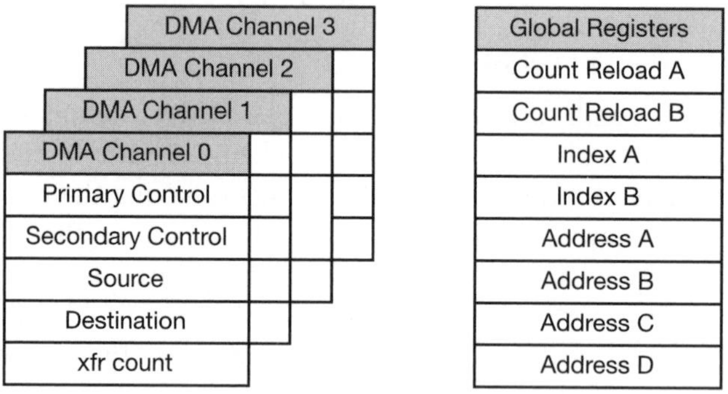

DMA Channel 3
DMA Channel 2
DMA Channel 1
DMA Channel 0
Primary Control
Secondary Control
Source
Destination
xfr count

Global Registers
Count Reload A
Count Reload B
Index A
Index B
Address A
Address B
Address C
Address D

Figure 8.10 DMA memory-mapped registers

Primary control register

This register controls the main operation of the DMA and contains 16 bit-fields as shown in Figure 8.11. Refer to the TMS320C6201/C6701 *Peripherals Reference Guide* (SPRU190) for the bit-field description.

Secondary control register

This register holds eight sets of conditions (COND) and interrupt enable (IE) bit-fields for monitoring events and validating interrupt to the CPU respectively (refer to the User Guide).

Source and destination registers

The source and destination registers (32-bit wide) hold the address of the next read and write transfers, respectively.

Transfer counter register

This register contains two separate sets of 16 bit-fields that hold the number of frames and elements per frame to be transferred.

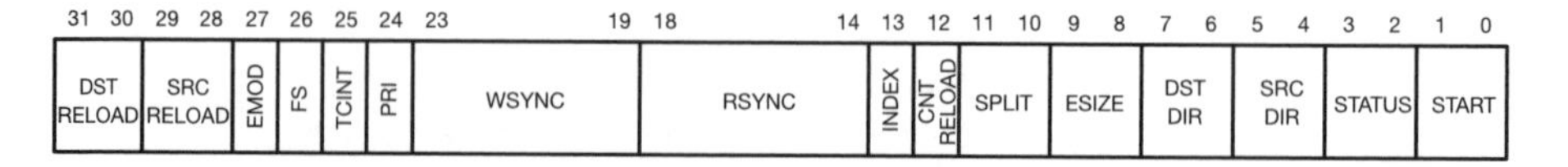

Figure 8.11 DMA primary control register

Count reload registers

There are two count reload registers (register A and register B) which hold the number of frames and the number of elements per frame to be reloaded into the counter register. This is controlled by the CNT RELOAD field of the Primary Control Register.

DMA global address registers A, B, C and D

The four global address registers A, B, C and D can be used as a source or destination address reload for auto-initialisation. This is useful when one DMA channel can be used for transferring data to/from different locations (see Program 8.7 below).

Index registers A and B

These two registers can be used for holding the frame and element indexes when multi-frame transfer is used.

8.6 Practical example for programming the DMA

In this example, the DMA channel 2 is programmed to transfer a block (`BUFFER_SIZE`) of data from a fixed source address (Serial Port zero) to a buffer (InBuffer1 or InBuffer2). This DMA channel is programmed such that

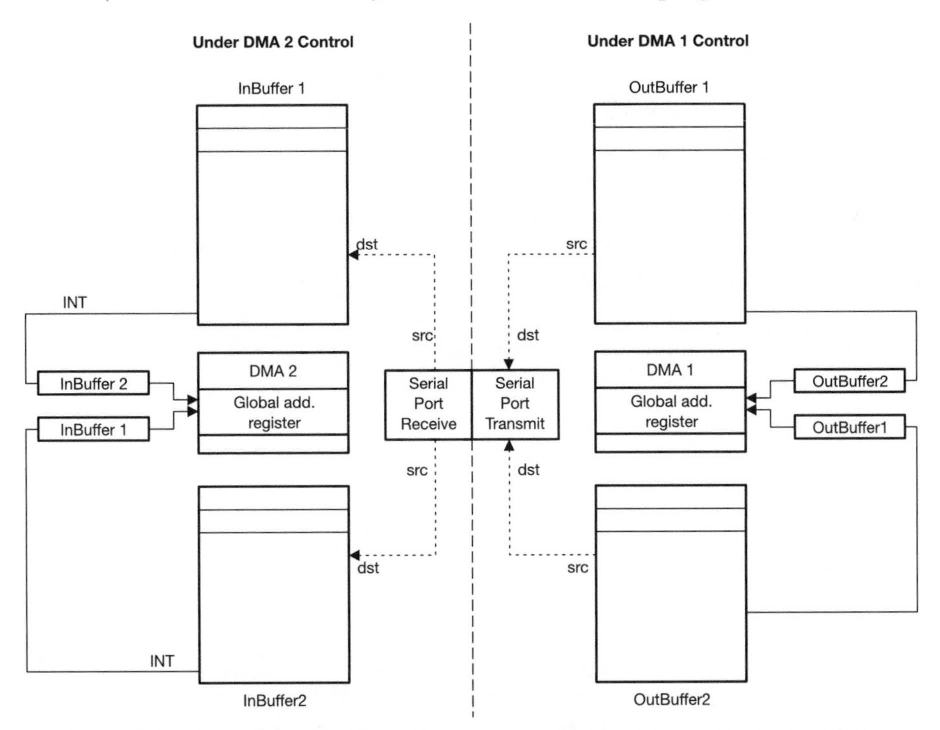

Figure 8.12 Illustration of DMA channel 1 and DMA channel 2 operations

when a buffer is full it generates a DMA interrupt to the CPU. The interrupt service routine then performs the Goertzel algorithm and reloads the Global Address Register B with the next buffer (InBuffer2 or InBuffer1); this ensures that the destination address is changed from InBuffer1 to InBuffer2 to InBuffer1, etc. However, the DMA channel 1 is programmed to shift the processed data from a buffer (OutBuffer1 or OutBuffer2) to the Serial Port zero. The same interrupt service routine also reloads the Global Address Register C with the next buffer (OutBuffer1 or OutBuffer2) so that the source address is changed from OutBuffer1 to OutBuffer2 to OutBuffer1, etc. This is illustrated in Figure 8.12 and Program 8.7. The complete program can be found in directory `F:\DSPCODE\Goertzel (Chap8)\GTZDMA\`.

```
/*File:       gtzdma.c
Description: Perform goertzel algorithm on 200 samples of codec input.
             Sends goertzel value periodically to mailbox. Also sends the original sample modulated with the
             goertzel value out to both channels of the codec. The program also uses double buffering techniques
             and auto-initialising dma to send and receive codec data.
-----------------------------------------------------------------------*/

#include <stdio.h>
#include <string.h>
#include <common.h>
#include <codec.h>
#include <mcbspdrv.h>
#include <board.h>
#include <stdlib.h>
#include <timer.h>
#include <intr.h>
#include <regs.h>
#include <dma.h>
#include <pci.h>

#define SAMPLING_RATE 20000

extern int inicodec (int sampling_rate);

#pragma CODE_SECTION(DMA_Interrupt,".iprog")
#pragma DATA_SECTION(InBuffer1,".edata")
#pragma DATA_SECTION(InBuffer2,".edata")
#pragma DATA_SECTION(OutBuffer1,".edata")
#pragma DATA_SECTION(OutBuffer2,".edata")
#pragma DATA_SECTION(ptrInBuffer,".edata")
#pragma DATA_SECTION(ptrOutBuffer,".edata")

#define BUFFER_SIZE 200
unsigned int InBuffer1 [BUFFER_SIZE];       // First of 2 alternate input buffer to store serial data.
unsigned int InBuffer2 [BUFFER_SIZE];       // Second of 2 alternate input buffer to store serial data.
unsigned int *ptrInBuffer;                  // Pointer (either == InBuffer1 or == InBuffer2) waiting to be filled.
unsigned int OutBuffer1 [BUFFER_SIZE];      // First of 2 alternate output buffer to send serial data.
unsigned int OutBuffer2 [BUFFER_SIZE];      // Second of 2 alternate output buffer to send serial data.
unsigned int *ptrOutBuffer;                 // Pointer (either == OutBuffer1 or == OutBuffer2) yet to have data to send.
/* -----------------------------------------------------------------------
   Configure DMAs to perform autobuffering alternating between 2 buffers
------------------------------------------------------------------------- */
inline void Init_DMA (void)
{
  DMA_GCR_A = BUFFER_SIZE;                  // Reload counter after block transfer.

  //=============== DMA 2 for receiving codec data ========================
  DMA_GADDR_B = (unsigned int) InBuffer2;   // Set next buffer to 2nd alternate buffer.
  DMA2_PRIMARY_CTRL =

      (DMA_RELOAD_GARB << 30)               // Reload destination address with global address B.
    | (DMA_RELOAD_NONE << 28)               // Don't reload source address (fixed)
    | (0               << 27)               // Emode, DMA halts(1) continues(0) during emulation.
    | (0               << 26)               // FS=0, No frame sync.
    | (1               << 25)               // TCINT=1, enable dma interrupt.
```

Program 8.7 Complete code for implementing the Goertzel algorithm using DMA

Program 8.7 continued

```
    | (DMA_DMA_PRI     << 24)                      // DMA has priority of CPU on internal bus.
    | (SEN_NONE        << 19)                      // No need Write Sync. write to buffer asap.
    | (SEN_REVT0       << 14)                      // Read Sync from receive-buffer-full of MCBSP0.
    | (DMA_GNDX_A      << 13)                      // Use global index register A as programmable index.
    | (DMA_CNT_RELOADA << 12)                      // Use global count reload register A for auto reload.
    | (DMA_SPLIT_DIS   << 10)                      // Disable split channel mode.
    | (DMA_ESIZE32     <<  8)                      // Set element size to 32-bit.
    | (DMA_ADDR_INC    <<  6)                      // Auto-increment destination address after each transfer.
    | (DMA_ADDR_NO_MOD <<  4)                      // Fix source address.
    | (DMA_STOP_VAL         );                     // Stop dma now. Start later, after configuration.
  DMA2_SECONDARY_CTRL =
      (1               << 15)                      // Clear any prior WSYNC status.
    | (1               << 13)                      // Clear any prior RSYNC status.
    | (1               <<  7);                     // Enable block-transfer-end interrupt.
  DMA2_SRC_ADDR = (unsigned int) MCBSP_DRR_ADDR (0); // Source address is receive data of MCBSP0
  DMA2_DEST_ADDR = (unsigned int) InBuffer1;       // Destination address is 1st of 2 DMA buffers.
  DMA2_XFER_COUNTER = BUFFER_SIZE;                 // Set to transfer BUFFER_SIZE words of 32 bits each.
  ptrInBuffer = InBuffer1;                         // Set the current buffer to be filled.

  //================ DMA 1 for transmitting codec data =========================
  DMA_GADDR_C = (unsigned int) OutBuffer2;         // Set next buffer to 2nd alternate buffer.

  DMA1_PRIMARY_CTRL =
      (DMA_RELOAD_NONE << 30)                      // Don't reload destination address (fixed)
    | (DMA_RELOAD_GARC << 28)                      // Reload source address with global address C.
    | (0               << 27)                      // Emode, DMA halts(1) continues(0) during emulation.
    | (0               << 26)                      // FS=0, No frame sync.
    | (0               << 25)                      // TCINT=1 => enable dma interrupt.
    | (DMA_DMA_PRI     << 24)                      // DMA has priority of CPU on internal bus.
    | (SEN_XEVT0       << 19)                      // Write Sync. from transmit buffer of MCBSP0.
    | (SEN_NONE        << 14)                      // No need Read Sync.
    | (DMA_GNDX_A      << 13)                      // Use global index register A as programmable index.
    | (DMA_CNT_RELOADA << 12)                      // Use global count reload register A for auto reload.
    | (DMA_SPLIT_DIS   << 10)                      // Disable split channel mode.
    | (DMA_ESIZE32     <<  8)                      // Set element size to 32-bit.
    | (DMA_ADDR_NO_MOD <<  6)                      // Fixed destination.
    | (DMA_ADDR_INC    <<  4)                      // Auto-increment source address.
    | (DMA_STOP_VAL         );                     // Stop dma now. Start later, after configuration.

  DMA1_SECONDARY_CTRL =
      (1               << 15)                      // Clear any prior WSYNC status.
    | (1               << 13);                     // Clear any prior RSYNC status.

  DMA1_DEST_ADDR = (unsigned int) MCBSP_DXR_ADDR (0); // Source address is receive data of MCBSP0
  DMA1_SRC_ADDR = (unsigned int) OutBuffer1;       // Destination address is 1st of 2 DMA buffers.
  DMA1_XFER_COUNTER = BUFFER_SIZE;                 // Set to transfer BUFFER_SIZE words of 32 bits each.
  ptrOutBuffer = OutBuffer1;                       // Set the current buffer to be filled.

  /*************************************************************************
    Note: DMA2 interrupt to process data.
          In this implementation, channel 1 is started first and it has
          higher priority than channel 2.
          In this way, we are sure that by the time we process the data
          in DMA2 interrupt, both reload addresses have been used.
  ************************************************************************/
  memset (OutBuffer1, 0, BUFFER_SIZE * sizeof (int));
  memset (OutBuffer2, 0, BUFFER_SIZE * sizeof (int));
  DMA_AUTO_START (1);                              // Start DMA1 with auto-reload.
  DMA_AUTO_START (2);                              // Start DMA2 with auto-reload.

  return;
}

interrupt void DMA_Interrupt (void)
{
  short delay_2 = 0;
  short delay_1 = 0;
  short delay;
  int input;
  short coef_1 = 0x79bc;                           // 0.9511 in decimal. (1kHz @ 20kHz)
  int i;
  int Goertzel_Value;
  int prod1, prod2, prod3;

  //============== evaluate goertzel value ==================
  //          attenuate input to avoid overflow
  //=========================================================

  for (i=0; i<(BUFFER_SIZE); i++)
```

Program 8.7 continued

```
  {
    input = ptrInBuffer[i]<<16;
    input >>= 22;
    prodl = delay_1 * coef_1;
    prodl >>= 14;
    delay = (short) input + (short) prodl - delay_2;
    delay_2 = delay_1;
    delay_1 = delay;
  }
  prodl = ((delay_1) * (delay_1)) >> 15;
  prod2 = ((delay_2) * (delay_2)) >> 15;
  prod3 = ((delay_1) * (delay_2)) >> 15;
  prod3 = (prod3 * coef_1) >> 14;
  Goertzel_Value = prodl + prod2 -- prod3;
  if (Goertzel_Value <0) Goertzel_Value = 0;          // Value cannot be negative. (Magnitude)
  *(unsigned int *)(PCI_MAP1_BASE_ADDR+AMCC_AOMB1_OFFSET)
    = (unsigned int) Goertzel_Value;                  // Send value out to mailbox.

  //============ modulate sample with goertzel value =============

  for (i=0; i<BUFFER_SIZE; i++)
  {
    input = ptrInBuffer[i]>>16;
    prodl = ((short) input * (short)(Goertzel_Value)) >> 15;
    prodl = prodl & 0xffff;
    prodl = (prodl<<16) | prodl;
    ptrOutBuffer[i] = prodl;
  }

  //============ send processed sample =============
  DMA_GADDR_B = (unsigned int) ptrInBuffer;           // Prepare reload address for next block.
  DMA_GADDR_C = (unsigned int) ptrOutBuffer;          // Prepare reload address for next block.
  if (ptrInBuffer == InBufferl)                       // Alternate the ptrCurrentBuffer between the
  {
    ptrInBuffer = InBuffer2;
    ptrOutBuffer = OutBuffer2;
  }
  else
  {
    ptrInBuffer = InBufferl;
    ptrOutBuffer = OutBufferl;
  }
  DMA2_SECONDARY_CTRL &= 0xffffffbf;                  // Clear block-transfer-end condition.
}

void Init_Interrupts ()
{
  intr_reset();

  INTR_CLR_FLAG (CPU_INT9);
  intr_hook (DMA_Interrupt, CPU_INT9);
  intr_map (CPU_INT9, ISN_DMA_INT2);
  INTR_ENABLE (CPU_INT9);

}

void main(void)
{
  evm_init();                                         // Initialise EVM board.
  Init_DMA ();
  inicodec (SAMPLING_RATE);                           // Initialise codec, adjust sampling rate.
  Init_Interrupts ();                                 // Initialise interrupts and hook isr.
  INTR_GLOBAL_ENABLE();                               // Enable global interrupt.
  for (;;);                                           // Main loop, does nothing.
}
```

Chapter 9

Implementation of the Discrete Cosine Transform

9.1 Introduction

Due to the increasing need of video compression for storage and transmission, many international standards have emerged. Different standards exist for different applications. For instance, the JPEG (Joint Photographic Experts Group) standard (ISO standard 10918) has been designed for coding still images, whereas the H.261 standard which is part of the H.320 standard is designed for video-conferencing, and the MPEG1, MPEG2 and MPEG4 (Moving Pictures Experts Group) standards (MPEG7 is currently under development) are designed to support more advanced video communication applications.

The JPEG standard was designed in 1992 for compression and recovery of still photographic images. The standard provides a relatively good compression ratio and is less computationally intensive than the MPEG standard, which was developed for transmission of moving pictures. The JPEG standard is centred on the block-based Discrete Cosine Transform (DCT).

Different modes (sequential, progressive and hierarchical) and options (lossy, lossless, arithmetical, etc.) of the JPEG standard exist. To perform the JPEG coding, an image (in colour or grey scales) is first sub-divided into blocks of 8×8 pixels and then the DCT is performed on each block. The DCT operation then generates 64 coefficients, which are quantised to reduce the magnitude of the coefficients (see quantisation matrix in Program 9.1 below) and reordered into a one-dimensional array in a zigzag manner

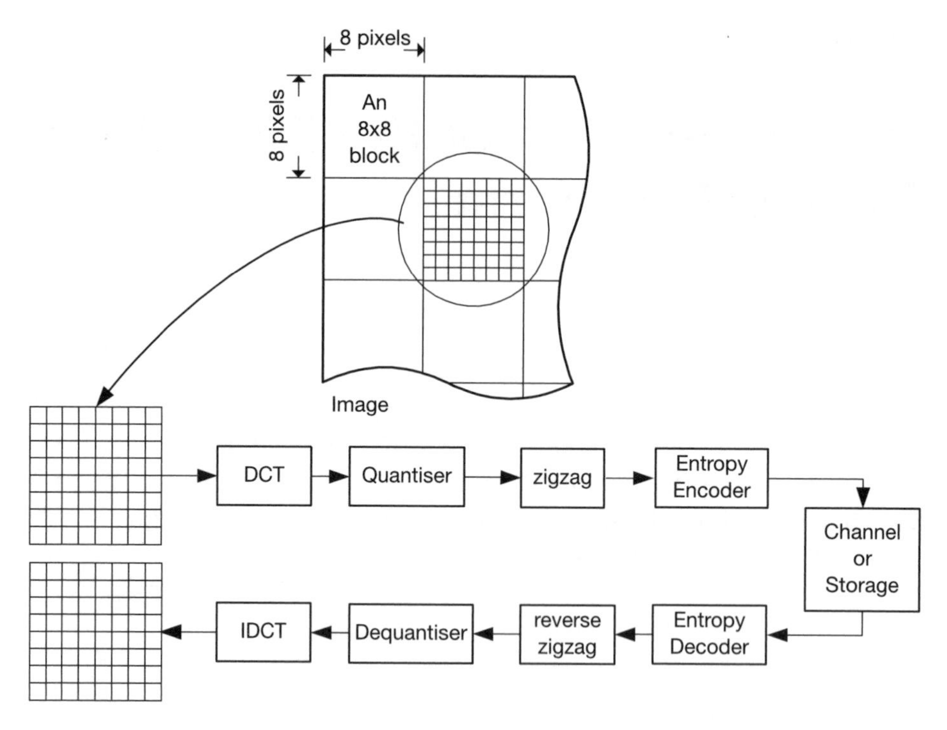

Figure 9.1 JPEG encoder–decoder sequence

before further encoding. The compression is achieved at different stages; the first one is during quantisation and the second is during the entropy coding process.

To perform JPEG decoding the reverse process of coding is applied, that is the encoded data stream is entropy decoded and passed through a reverse zigzag table in order to generate a 2-D array. This array is then rescaled using the quantising factor and then passed through the Inverse DCT (IDCT) (see Figure 9.1). For more details on the JPEG process, see Mattison (1994). For simplicity only the DCT and IDCT will be dealt with. PC–DSP communication is discussed and used for displaying the processed images.

There are a number of transforms that can be used for image and video coding, such as the Walsh–Hadamard transform and the Karhunen–Loeve transform. However, since its introduction in 1974 (Ahmed *et al.*, 1974), the DCT has become the most popular transform in image and video coding. Today, it constitutes the heart of almost all image and video coding standards. This popularity is due to the fact that the DCT provides the best compromise between energy compaction and computational complexity.

DCT-based codecs use a two-dimensional version of the transform. The two-dimensional DCT and its inverse (IDCT) of an $N \times N$ block of pixels are shown in Equations [9.1] and [9.2] respectively. Note that the DCT, in a sense, is similar to the Discrete Fourier Transform (DFT) since it decomposes a signal into a series of harmonic cosine functions.

2-D DCT:

$$F(u,v) = \frac{2}{N} C(u)C(v) \sum_{y=0}^{N-1} \sum_{x=0}^{N-1} f(x,y) \cos\left[\frac{(2x+1)u\pi}{2N}\right] \cos\left[\frac{(2y+1)v\pi}{2N}\right] \quad [9.1]$$

2-D IDCT:

$$f(x,y) = \frac{2}{N} \sum_{v=0}^{N-1} \sum_{u=0}^{N-1} C(u)C(v)F(u,v) \cos\left[\frac{(2x+1)u\pi}{2N}\right] \cos\left[\frac{(2y+1)v\pi}{2N}\right] \quad [9.2]$$

where

$$C(a) = \begin{cases} \frac{1}{\sqrt{2}}, & a = 0 \\ 1, & \text{otherwise} \end{cases}$$

$f(x,y)$ is the pixel intensity at x and y coordinates, and
$F(u,v)$ is its corresponding DCT coefficient.

From Equation [9.1], it can be deduced that we need N^4 multiplications to calculate the DCT of an $N \times N$ block. Thus, to calculate the DCT of a 256×256 image we need $256^4 = 4{,}294{,}967{,}296$ multiplications. However, if we break this image into 1024 blocks of 8×8 pixels, then we need only $1024 \times 8^4 = 4{,}194{,}304$ multiplications. As noted previously, the goal of transform coding is to concentrate energy amongst as few coefficients as possible. As an image will normally contain many different frequencies, the energy compaction task becomes very difficult. If, however, the image is broken into blocks, then each block will contain only a small subset of similar frequencies, and energy compaction becomes easy and more efficient.

The one-dimensional DCT and IDCT can be implemented as shown in Equations [9.3] and [9.4]. It is noticeable from Equations [9.1], [9.2], [9.3] and [9.4] that the DCT and IDCT coefficients are constant and therefore can be pre-calculated and used prior to the transform in order to save processing time. The coefficients are shown in Q12 format in Figure 9.2. An example of a two-dimensional DCT is shown in Figure 9.3. These figures are discussed in Section 9.2.1.

$$\text{DCT:} \quad X(k) = \sqrt{\frac{2}{N}} C(k) \sum_{i=0}^{N-1} x(i) \cos\left[\frac{(2i+1)k\pi}{2N}\right] \quad [9.3]$$

$$\text{IDCT:} \quad x(n) = \sqrt{\frac{2}{N}} \sum_{u=0}^{N-1} C(u)X(u) \cos\left[\frac{(2u+1)n\pi}{2N}\right] \quad [9.4]$$

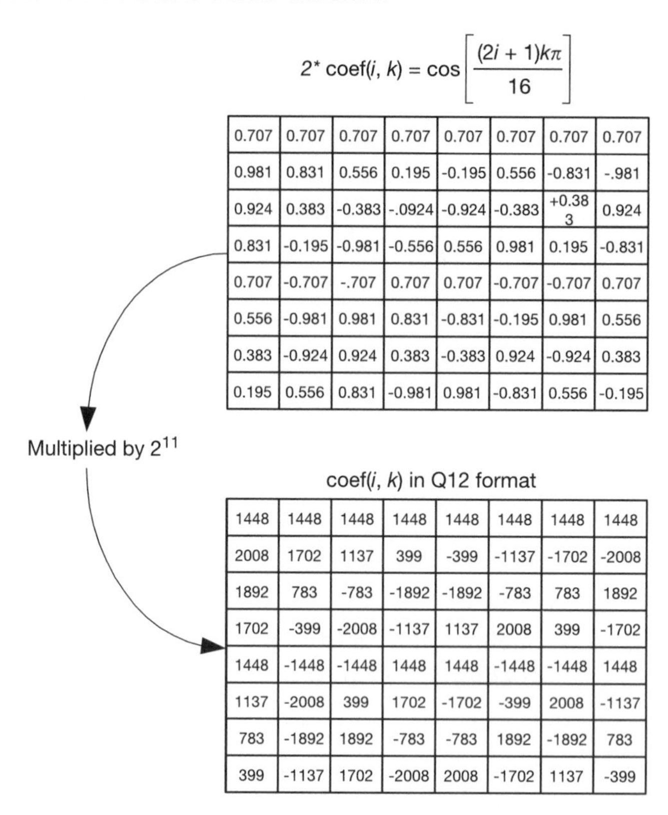

$$2^* \text{coef}(i, k) = \cos\left[\frac{(2i+1)k\pi}{16}\right]$$

0.707	0.707	0.707	0.707	0.707	0.707	0.707	0.707
0.981	0.831	0.556	0.195	-0.195	0.556	-0.831	-.981
0.924	0.383	-0.383	-.0924	-0.924	-0.383	+0.383	0.924
0.831	-0.195	-0.981	-0.556	0.556	0.981	0.195	-0.831
0.707	-0.707	-.707	0.707	0.707	-0.707	-0.707	0.707
0.556	-0.981	0.981	0.831	-0.831	-0.195	0.981	0.556
0.383	-0.924	0.924	0.383	-0.383	0.924	-0.924	0.383
0.195	0.556	0.831	-0.981	0.981	-0.831	0.556	-0.195

Multiplied by 2^{11}

coef(i, k) in Q12 format

1448	1448	1448	1448	1448	1448	1448	1448
2008	1702	1137	399	-399	-1137	-1702	-2008
1892	783	-783	-1892	-1892	-783	783	1892
1702	-399	-2008	-1137	1137	2008	399	-1702
1448	-1448	-1448	1448	1448	-1448	-1448	1448
1137	-2008	399	1702	-1702	-399	2008	-1137
783	-1892	1892	-783	-783	1892	-1892	783
399	-1137	1702	-2008	2008	-1702	1137	-399

Figure 9.2 DCT coefficients in decimal and Q12 format

9.2 Optimisation of DCT and IDCT for DSP implementation

As previously discussed, to optimise any algorithm for DSP implementation, we must first analyse the algorithm itself and try to reduce its complexity, at the algorithmic level, before we proceed to the next step. In the case of optimising the DCT and the IDCT this will prove to be very invaluable. The DCT and IDCT algorithm optimisation will be achieved in two stages. In the first stage, the 2-D DCT will be separated into a 1-D DCT pair. In the second stage, the 1-D DCT pair will be further optimised using the McGovern fast 1-D DCT.

9.2.1 Two-dimensional DCT using a one-dimensional DCT pair

One of the properties of the 2-D DCT is that it is separable. This means that it can be separated into a pair of 1-D DCTs. Thus, to obtain the 2-D DCT of a

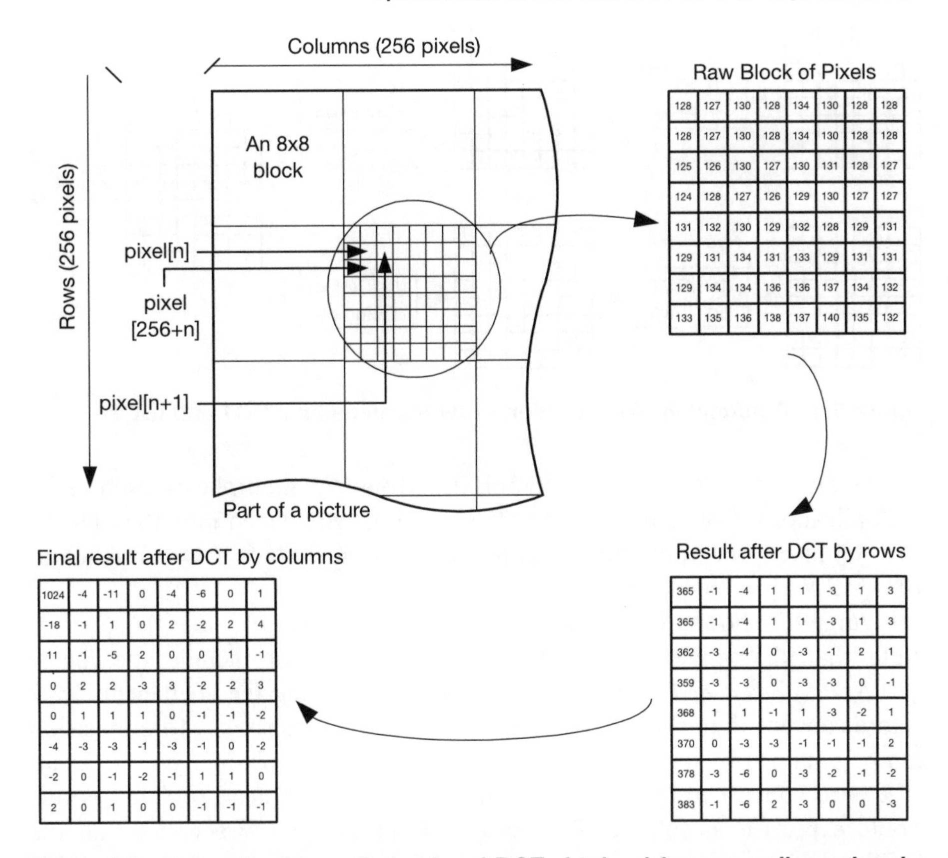

Raw Block of Pixels

128	127	130	128	134	130	128	128
128	127	130	128	134	130	128	128
125	126	130	127	130	131	128	127
124	128	127	126	129	130	127	127
131	132	130	129	132	128	129	131
129	131	134	131	133	129	131	131
129	134	134	136	136	137	134	132
133	135	136	138	137	140	135	132

Result after DCT by rows

365	-1	-4	1	1	-3	1	3
365	-1	-4	1	1	-3	1	3
362	-3	-4	0	-3	-1	2	1
359	-3	-3	0	-3	-3	0	-1
368	1	1	-1	1	-3	-2	1
370	0	-3	-3	-1	-1	-1	2
378	-3	-6	0	-3	-2	-1	-2
383	-1	-6	2	-3	0	0	-3

Final result after DCT by columns

1024	-4	-11	0	-4	-6	0	1
-18	-1	1	0	2	-2	2	4
11	-1	-5	2	0	0	1	-1
0	2	2	-3	3	-2	-2	3
0	1	1	1	0	-1	-1	-2
-4	-3	-3	-1	-3	-1	0	-2
-2	0	-1	-2	-1	1	1	0
2	0	1	0	0	-1	-1	-1

Figure 9.3 Example of two-dimensional DCT obtained from one-dimensional DCT

block, a 1-D DCT is first performed on the rows of the block and then a 1-D DCT is performed on the columns of the resulting block. The same applies to the IDCT. This process is illustrated in Figure 9.4. A point to note here is that the order in which the 1-D transforms are applied (i.e. rows then columns or columns then rows) does not affect the final result. The 1-D DCT and IDCT are given in Equations [9.5] and [9.6].

$$\text{1-D DCT:} \qquad F(u) = \sqrt{\frac{2}{N}} C(u) \sum_{x=0}^{N-1} f(x) \cos\left[\frac{(2x+1)u\pi}{2N}\right] \qquad [9.5]$$

$$\text{1-D IDCT:} \qquad f(x) = \sqrt{\frac{2}{N}} \sum_{u=0}^{N-1} C(u) F(u) \cos\left[\frac{(2x+1)u\pi}{2N}\right] \qquad [9.6]$$

From Equation [9.5], it can be deduced that to calculate the 2-D DCT of an $N \times N$ block we need $2N^3$ multiplications. Thus, separating the 2-D DCT into

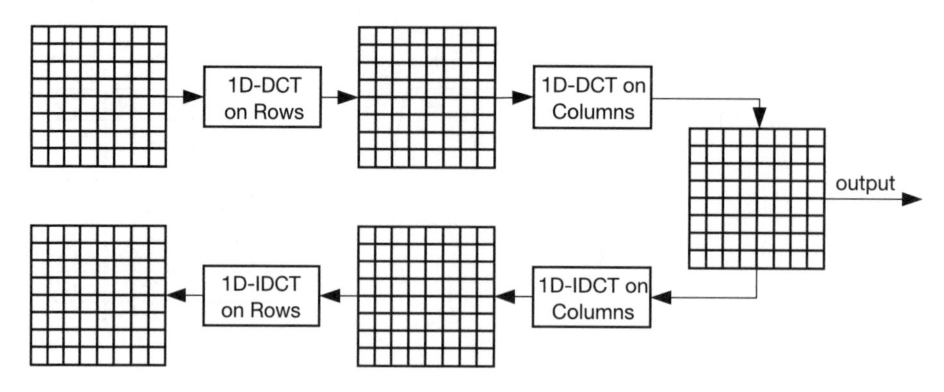

Figure 9.4 Practical implementation of two-dimensional DCT and IDCT

a 1-D DCT pair reduces the complexity from N^4 multiplications to $2N^3$ multiplications. For example, for a 256×256 image divided into 1024 blocks of 8×8 pixels, the multiplications are reduced from $1024 \times 8^4 = 4{,}194{,}304$ to $1024 \times 2 \times 8^3 = 1{,}048{,}576$.

Figure 9.2 shows the coefficients both in decimal and in Q12 format. Note that the values in the decimal format do not include the factor $\sqrt{2/N} = \sqrt{2/8} = \frac{1}{2}$. The reason for this is that dividing the coefficients by 2 before converting to Q12 may result in some loss in precision. More precision can be obtained by ignoring this division and then multiplying by 2^{11} (instead of 2^{12}) to convert to Q12. The reason for using Q12 instead of Q15 (as one would expect) is as follows. Referring to Equation [9.5], we notice that the DCT calculation involves a summation of N terms. Looking at the DCT coefficients in Figure 9.2, we observe that the terms entering in the summation may be close to 1. Thus, such summation may cause overflows. To avoid this, each term must be scaled down by $1/N$. For $N = 8$, this can be achieved by working in Q12 format instead of Q15 format.

An example of a 2-D DCT applied to an 8×8 block of pixels has been shown in Figure 9.3. Note that most of the energy is now concentrated in the low frequency coefficients (especially the DC coefficient), and that most high frequency coefficients are zero or near zero.

9.3 Block-based DCT and IDCT in C

The most straightforward way of implementing a DCT and an IDCT is to use the one-dimensional DCT and IDCT as explained above. Program 9.1 shows how to translate Equations [9.5] and [9.6] into C code. The program also takes into account the numerical issues associated with fixed-point processors. The complete program can be found in directory `F:\DSPCODE\DCT (Chap9)\SLOWDCT\`.

```
/***********************************************/
/* DCT and IDCT functions */
/***********************************************/
inline void dct(void)
{
  int i,j,x,y;
  int value[8];

  /* perform 1D DCT on the rows */
  for(i=0;i<64;i+=8)
  {
    for(y=0;y<8;++y)
    {
      value[y] = 0;

      for(x=0;x<8;++x)
        value[y] += (coe[y][x]*block[i+x]);
    }
    for(y=0;y<8;++y)
      block[i+y] = (value[y]>>12);
  }

  /* perform 1D DCT on the columns */
  for(j=0;j<8;j++)
  {
    for(y=0;y<8;++y)
    {
      value[y]=0;

      for(x=0;x<8;++x)
        value[y] += (coe[y][x]*block[j+(x*8)]);
    }
    for(y=0;y<8;++y)
      block[j+(y*8)] = (value[y]>>12);
  }
}

inline void idct(void)
{
  int i,j,x,y;
  int value[8];

  /* perform 1D IDCT on the rows */
  for(i=0;i<64;i+=8)
  {
    for(y=0;y<8;++y)
    {
      value[y] = 0;
      for(x=0;x<8;++x)
        value[y] += (int)(coe[x][y]*block[i+x]);
    }
    for(y=0;y<8;++y)
      block[i+y] = (short)(value[y]>>12);
  }
```

Program 9.1 DCT and IDCT in C

Program 9.1 continued

```
  /* perform 1D IDCT on the columns */
  for(j=0;j<8;j++)
  {
    for(y=0;y<8;++y)
    {
      value[y] = 0;
      for(x=0;x<8;++x)
        value[y] += (int)(coe[x][y]*block[j+(x*8)]);
    }
    for(y=0;y<8;++y)
      block[j+(y*8)] = (short)(value[y]>>12);
  }

  for(i=0;i<64;i++)
  {
    if (block[i] < 0) block[i] = 0;
    if (block[i] > 255) block[i] = 255;
  }
}
```

9.4 Simple DSP implementation of DCT and IDCT on an image

In Program 9.1 the DCT and IDCT functions process an 8×8 block of an image. In practical situations the DCT and IDCT are performed on the whole image. Therefore Program 9.1 will be slightly modified. Let us first look at how a picture is composed and how a block within an image can be accessed. An image can be subdivided into blocks of 8×8 pixels and each block can be represented by a row and a column. Yet the image is stored in a one-dimensional array. Therefore in order to extract a block from the image array, the sequence of code shown in Program 9.2 could be used.

```
for (y=0;y<8;y++)
{
  for (x=0;x<8;x++)
  {
    i = image_in[(col*8+x)+(row*8+y)*256];  /* col and row represent the blocks coordinates */
    *pTmp++ = (int) i;
  }
}
```

Program 9.2 Block extraction from an image array

Therefore by selecting a row and column and using Program 9.2 a block within an image can be extracted. The C code shown in Program 9.3

demonstrates a complete implementation of 2-D DCT and 2-D IDCT using 1-D DCT and 1-D IDCT. The program performs DCT and IDCT on 8×8 blocks that are read from an image (`scenary.h`) located in the on-board memory. The programs are self-explanatory, though a few clarifications may prove to be useful.

- The program runs on an EVM and does not use any communication with the host.
- The `scenary.h` file is quite large (256×256 8-bit pixels, or 65,536 bytes) and therefore needs to be located in a specific memory location in the EVM. By using the `#pragma DATA_SECTION()` directive, the `scenary.h` file could be located in the `SDRAM0_DATA_MEM` (origin address 0x02000000): see command file `F:\DSPCODE\DCT (Chap9)\SLOWDCT\COM.CMD`.
- The `#pragma DATA_SECTION (image_in, 'myvar0')` allows the `image_in` data array to be loaded into the section called `myvar0` which is defined in the command file. Similarly the `image_out` array which contains the output of the IDCT is stored in the section called `SDRAM1_DATA_MEM` (origin 0x0300 0000) using the same principle.
- The directory `F:\DSPCODE\DCT (Chap9)\SLOWDCT\` contains the entire source and executable files required. The program can be loaded into the EVM and executed, and by using the command '`mem1 image_in`' and '`mem2 image_out`' (within the command line of the EVM debugger), the memory contents of `image_in` and `image_out` can be displayed and compared (see Appendix B).
- Table 9.1 shows a brief description of the programs used.

The following command line is used for compiling, assembling and linking:

```
cl6x -gs -als  dct_main.c dct.c idct.c  -z com.cmd
```

`dct_main.c`	This is the main file; it calls the DCT and IDCT files.
`dct_main.h`	This is the header file.
`dct.c`	This is the file which performs the DCT.
`idct.c`	This is the file which performs the IDCT.
`scenary.h`	This is the file which contains the image.
`com.cmd`	This is the command file used by the linker.
`comlink.bat`	This is the batch file for compiling and linking the files.

Table 9.1 Brief program description

```
/*********************************************************************/
/* dct_main perfomes a dct and idct function on an image.            */
/*********************************************************************/

/***************************************************/
/* include files                                   */
/***************************************************/
#include <stdio.h>
#include <stdlib.h>

#include "dct_main.h"              /* Includes and Constants used */
#pragma DATA_SECTION (image_in,"myvar0")
#pragma DATA_SECTION (image_out,"myvar1")
#include "scenary.h"               /* an h file containing input image as a 1D array */

/* 1D array to hold output image */
unsigned char image_out[IMAGE_SIZE];

/* 1D array to hold the current block */
short block[BLOCK_SIZE];

/*********************************************************************/
/* Q12 DCT coefficients (actual coefficient x 2^12 )                 */
/*                                                                   */
/*********************************************************************/
const short coe[8][8]=
  {
    4096,   4096,   4096,   4096,   4096,   4096,   4096,  4096,
    5681,   4816,   3218,   1130,  -1130,  -3218,  -4816, -5681,
    5352,   2217,  -2217,  -5352,  -5352,  -2217,   2217,  5352,
    4816,  -1130,  -5681,  -3218,   3218,   5681,   1130, -4816,
    4096,  -4096,  -4096,   4096,   4096,  -4096,  -4096,  4096,
    3218,  -5681,   1130,   4816,  -4816,  -1130,   5681, -3218,
    2217,  -5352,   5352,  -2217,  -2217,   5352,  -5352,  2217,
    1130,  -3218,   4816,  -5681,   5681,  -4816,   3218, -1130
  };

/*********************************************************************/
/* FUNCTIONS USED                                                    */
/*********************************************************************/
/* dct.c */
void dct(void);

/* idct.c */
void idct(void);

/*********************************************************************/
/* MAIN FUNCTION                                                     */
/*********************************************************************/
void main()
{
  int row, col, x, y, i;

  /*************************************************
   *          FORWARD DCT/ INVERSE DCT             *
   *************************************************/
```

(a) dct_main.c

Program 9.3 Source code files required for running 2-D DCT and 2-D IDCT

Program 9.3a continued

```
  /* block by block processing */
  for (row=0; row<IMAGE_LEN; row+=BLOCK_LEN)
  {
    for (col=0; col<IMAGE_LEN; col+=BLOCK_LEN)
    {
      for (y=0, i=0; y<BLOCK_LEN; y++)       /* get the block from the input image */
      {
        for (x=0; x<BLOCK_LEN; x++, i++)
           block[i] = (short) image_in[(col+x)+(row+y)*IMAGE_LEN];
      }
      dct();                                 /* perform FDCT on this block */
      idct();                                /* perform IDCT on this block */
      for (y=0, i=0; y<BLOCK_LEN; y++)       /* store block to output image */
      {
        for (x=0; x<BLOCK_LEN; x++, i++)
        {
          if(block[i]<0)
            /* Quick fix for errors occuring due to negative a values occuring after IDCT!*/
            image_out[(col+x)+(row+y)*IMAGE_LEN]=(unsigned char) (-block[i]);
          else
            image_out[(col+x)+(row+y)*IMAGE_LEN]=(unsigned char) block[i];
        }
      }
    }
  }
  for (;;);                                  /* wait */
}
/***************************************************************************/
```

```
/*******************************************************/
/* dct.c Function to perform an 8 point 2D DCT         */
/* DCT is performed using direct matrix multiplication */
/*******************************************************/

/* externals */
#include "dct_main.h"

extern unsigned char image_in[IMAGE_SIZE];
extern unsigned char image_out[IMAGE_SIZE];
extern short block[BLOCK_SIZE];
extern const short coe[8][8];

/************************************************
 DCT and IDCT functions
************************************************/
void dct(void)
{
  int i,j,x,y;
  int value[8];

  /* perform 1D DCT on the rows */
  for(i=0;i<64;i+=8)
  {
    for(y=0;y<8;++y)
```

(b) dct.c

Program 9.3b continued

```
    {
      value[y] = 0;      for(x=0;x<8;++x)
        value[y] += (coe[y][x]*block[i+x]);
    }

    for(y=0;y<8;++y)
      block[i+y] = (value[y]>>12);
  }

  /* perform 1D DCT on the columns */
  for(j=0;j<8;j++)
  {
    for(y=0;y<8;++y)
    {
      value[y]=0;

      for(x=0;x<8;++x)
        value[y] += (coe[y][x]*block[j+(x*8)]);
    }

    for(y=0;y<8;++y)
      block[j+(y*8)] = (value[y]>>12);
  }
}
```

```
/*****************************************************/
/* idct.c performs an 8 point 2D Inverse DCT function */
/*****************************************************/

/* externals */
#include "dct_main.h"

extern unsigned char image_in[IMAGE_SIZE];
extern unsigned char image_out[IMAGE_SIZE];
extern short block[BLOCK_SIZE];
extern const short coe[8][8];

void idct(void)
{
  int i,j,x,y;
  int value[8];

  /* perform 1D IDCT on the rows */
  for(i=0;i<64;i+=8)
  {
    for(y=0;y<8;++y)
    {
      value[y] = 0;
      for(x=0;x<8;++x)
        value[y] += (int)(coe[x][y]*block[i+x]);
    }
    for(y=0;y<8;++y)
      block[i+y] = (short)(value[y]>>12);
  }
```

(c) `idct.c`

Program 9.3c continued

```
  /* perform 1D IDCT on the columns */
  for(j=0;j<8;j++)
  {
    for(y=0;y<8;++y)
    {
      value[y] = 0;
      for(x=0;x<8;++x)
        value[y] += (int)(coe[x][y]*block[j+(x*8)]);
    }
    for(y=0;y<8;++y)
      block[j+(y*8)] = (short)(value[y]>>12);
  }

  for(i=0;i<64;i++)
  {
    if (block[i] < 0) block[i] = 0;
    if (block[i] > 255) block[i] = 255;
  }
}
```

```
unsigned char image_in[256*256] = {
72,  57,  55,  67,  79,  58,  52,  72,  96,  81,  72,  76,  93, 107,  91,  84,  95,  66,  65,
85,  74,  83,  72,  71,  60,  65,  50,  41,  46,  74,  54,  34,  35,  33,  32,  37,  35,  28,
30,  35,  56,  61,  62,  61,  59,  53,  56,  58,  ......57,  65, 142, 176, 174, 170, 164, 168,
172, 168, 165, 166, 164, 163, 165, 167, 167, 166, 164, 163, 162, 165, 166, 161, 160, 161, 132};
```

(d) scenary.h

```
dct_main.obj
dct.obj
idct.obj

-m dct.map
-heap 0x300000
-c

-o dct.out

-l c:\Evm6x\Dsp\Lib\Drivers\drv6x.lib
-l c:\Evm6x\Dsp\Lib\DevLib\dev6x.lib
-l rts6201.lib

/* linker command file for examples (MAP 1) */

MEMORY
{
  INT_PROG_MEM (RX)     : origin = 0x00000000 length = 0x00010000
  SBSRAM_PROG_MEM (RX)  : origin = 0x00400000 length = 0x00020000
  SBSRAM_DATA_MEM (RW)  : origin = 0x00420000 length = 0x00020000
  SDRAM0_DATA_MEM (RW)  : origin = 0x02000000 length = 0x00400000
  SDRAM1_DATA_MEM (RW)  : origin = 0x03000000 length = 0x00400000
  INT_DATA_MEM (RW)     : origin = 0x80000000 length = 0x00010000
}
```

(e) Command file com.cmd

Program 9.3e continued

```
SECTIONS
{
  .vec:    load = 0x00000000
  .text:   load = SBSRAM_PROG_MEM
  .const:  load = INT_DATA_MEM
  .bss:    load = INT_DATA_MEM
  .data:   load = INT_DATA_MEM
  .cinit   load = SBSRAM_PROG_MEM
  .pinit   load = INT_DATA_MEM
  .stack   load = INT_DATA_MEM
  .far     load = INT_DATA_MEM
  .sysmem  load = SDRAM0_DATA_MEM
  .cio     load = INT_DATA_MEM
  myvar0   load = SDRAM0_DATA_MEM

  myvar1   load = SDRAM1_DATA_MEM}
```

```
/*****************************************
* Constants used within the Program      *
*****************************************/

#ifndef DCT_MAIN_H

#define IMAGE_LEN 256
#define IMAGE_SIZE (IMAGE_LEN*IMAGE_LEN)

#define BLOCK_LEN 8
#define BLOCK_SIZE (BLOCK_LEN*BLOCK_LEN)

#endif
```

(f) Header file `Dct_Main.h`

9.4.1 McGovern Fast 1-D DCT and IDCT

In Section 9.2.1 it has been shown that the computational complexity of the 2-D DCT (and IDCT) can be reduced by separating it into a pair of 1-D DCTs (and IDCTs). This complexity can be further reduced by using a fast 1-D DCT. There are a number of fast 1-D DCTs (and IDCTs) reported in the literature. As an example, we have chosen to implement the algorithm reported by McGovern *et al.* (1994). By exploiting the redundancy in the DCT coefficients matrix and some matrix manipulations, McGovern reduces the complexity of the 2-D DCT applied to an 8×8 block to only 12 multiplications. For a detailed derivation of the algorithm, the reader is referred to Appendix A.

Program 9.4 shows the DSP implementation of the 2-D DCT (and IDCT) using the McGovern algorithm. Note that this program is very similar to Program 9.3. The only difference is that the `dct` and `idct` functions were changed such that they use the McGovern algorithm instead of Equations [9.3] and [9.4]. The two-dimensional array `coe[8][8]` in the `main` function of Program 9.3 has also been replaced in Program 9.4 with a one-dimensional array `coe[12]` that corresponds to the 12 multiplication coefficients of the McGovern algorithm. Note that McGovern starts his derivation with a scaled set of coefficients (scaled up by 8). The `dct` and `idct` take this into account and scale the results down by 8 (in some cases this is shown as a shift right by 3 and in other cases it is combined with the Q12-to-decimal conversion as a shift right by 15). In Program 9.4, we show only the functions `main`, `dct` and `idct`, as all other header files, command files and batch files are identical to those of Program 9.3. The source code can be found in the directory `F:\DSPCODE\DCT (Chap9)\FASTDCT\`.

It is suitable, at this point, to compare the computational complexity of the different methods of implementing the 2-D DCT. This is summarised in Table 9.2. Note the large savings in processing time provided by the McGovern algorithm, and remember that the intermediate stage of separating the 2-D DCT into a 1-D DCT pair made the use of McGovern possible. Table 9.3 shows the benchmark results obtained by the 'C6x EVM profiler.

Method	Multiplications for $N \times N$ block	Multiplications for 8×8 block	Multiplications for 256×256 image divided into 1024 blocks of 8×8 pixels
Direct implementation (Equation [9.1])	N^4	4096	4,194,304
A pair of 1-D DCTs (Equation [9.3])	$2N^3$	1024	1,048,576
McGovern algorithm (a pair)	—	24	24,576

Table 9.2 Computational complexity of different methods of implementing the 2-D DCT

Type of function	Profiler results without -o3 option (cycles)	Profiler results with -o3 option (cycles)
DCT (using McGovern algorithm)	11,339	4729
IDCT (using McGovern algorithm)	11,140	5329
DCT (using Equation [9.1])	112,136	15,023
IDCT (using Equation [9.2])	111,944	15,812

Table 9.3 Benchmark results

```
/*****************************************************/
/* dct.c Function to perform an 8 point 2D DCT       */
/* DCT is performed using direct matrix multiplication */
/*****************************************************/

/* externals */
#include "dct_main.h"
extern unsigned char image_in[IMAGE_SIZE];
extern unsigned char image_out[IMAGE_SIZE];
extern short block[BLOCK_SIZE];
extern const short coe[12];

void dct(void)
{
  short ADD[20];                          /* Table of the addition coefficients */
  int M[12];                              /* Table of the results of the multiplication */
  int postadd1,postadd2;
  int i,j;

  for(j=0;j<8;j++) /*Rows*/
    {
    /* first set of addtions */
    ADD[0]= (block[j]+block[56+j]);       /* x(0)+x(7) */
    ADD[1]= (block[24+j]+block[32+j]);    /* x(3)+x(4) */
    ADD[2]= (block[8+j]+block[48+j]);     /* x(1)+x(6) */
    ADD[3]= (block[16+j]+block[40+j]);    /* x(2)+x(5) */
    ADD[4]= (block[j]-block[56+j]);       /* x(0)+x(7) */
    ADD[5]= (block[48+j]-block[8+j]);     /* x(6)-x(1) */
    ADD[6]= (block[24+j]-block[32+j]);    /* x(3)-x(4) */
    ADD[7]= (block[16+j]-block[40+j]);    /* x(2)-x(5) */

    /* second set of additions, this is done so previous additions
       do not need to be repeated */
    ADD[8]= (ADD[0]+ADD[1]);
    ADD[9]= (ADD[0]-ADD[1]);
    ADD[10]=(ADD[2]+ADD[3]);
    ADD[11]=(ADD[2]-ADD[3]);
    ADD[12]=(ADD[4]+ADD[6]);
    ADD[13]=(ADD[5]+ADD[7]);
    ADD[14]=(ADD[9]+ADD[11]);
    ADD[15]=(ADD[4]+ADD[5]);
    ADD[16]=(ADD[12]+ADD[13]);
    ADD[17]=(ADD[6]+ADD[7]);
    ADD[18]=(ADD[8]+ADD[10]);
    ADD[19]=(ADD[8]-ADD[10]);

    /* Multiplications carried out, */
    M[0]= (int)(coe[0]*ADD[9]);
    M[1]= (int)(coe[1]*ADD[14]);
    M[2]= (int)(coe[2]*ADD[11]);
    M[3]= (int)(coe[3]*ADD[4]);
    M[4]= (int)(coe[4]*ADD[15]);
    M[5]= (int)(coe[5]*ADD[5]);
    M[6]= (int)(coe[6]*ADD[12]);
```

(a) dct.c

Program 9.4 Programs for implementing fast DCT and IDCT

Program 9.4a continued

```
    M[7]= (int)(coe[7]*ADD[16]);
    M[8]= (int)(coe[8]*ADD[13]);
    M[9]= (int)(coe[9]*ADD[6]);
    M[10]=(int)(coe[10]*ADD[17]);
    M[11]=(int)(coe[11]*ADD[7]);

    /* post multiplication, additions + subtractions */
    block[j]=ADD[18];                                  /* y(0) */
    block[32+j]=ADD[19];                               /* y(4) */
    block[16+j]=(short)((M[0]+M[1])>>12);              /* y(2) */
    block[48+j]=(short)((M[1]-M[2])>>12);              /* y(6) */
    postadd1= M[6]+M[7];
    postadd2= M[7]+M[8];
    block[56+j]= (short)((M[3]+M[4]+postadd1)>>12);    /* y(7) */
    block[40+j]= (short)((M[4]+M[5]+postadd2)>>12);    /* y(5) */
    block[8+j]= (short)((M[9]+M[10]-postadd1)>>12);    /* y(1) */
    block[24+j]= (short)((postadd2-M[10]-M[11])>>12);  /* y(3) */
  }

  for(i=0;i<64;i+=8)                                   /* Columns */
  {
    /* first set of additions */
    ADD[0]=(block[i]+block[i+7]);                      /* x(0)+x(7) */
    ADD[1]=(block[i+3]+block[i+4]);                    /* x(3)+x(4) */
    ADD[2]=(block[i+1]+block[i+6]);                    /* x(1)+x(6) */
    ADD[3]=(block[i+2]+block[i+5]);                    /* x(2)+x(5) */
    ADD[4]=(block[i]-block[i+7]);                      /* x(0)-x(7) */
    ADD[5]=(block[i+6]-block[i+1]);                    /* x(6)-x(1) */
    ADD[6]=(block[i+3]-block[i+4]);                    /* x(3)-x(4) */
    ADD[7]=(block[i+2]-block[i+5]);                    /* x(2)-x(5) */

    /* second set of additions, this is done so previous additions
       do not need to be repeated */
    ADD[8]= (ADD[0]+ADD[1]);
    ADD[9]= (ADD[0]-ADD[1]);
    ADD[10]=(ADD[2]+ADD[3]);
    ADD[11]=(ADD[2]-ADD[3]);
    ADD[12]=(ADD[4]+ADD[6]);
    ADD[13]=(ADD[5]+ADD[7]);
    ADD[14]=(ADD[9]+ADD[11]);
    ADD[15]=(ADD[4]+ADD[5]);
    ADD[16]=(ADD[12]+ADD[13]);
    ADD[17]=(ADD[6]+ADD[7]);
    ADD[18]=(ADD[8]+ADD[10]);
    ADD[19]=(ADD[8]-ADD[10]);

    /* Multiplications carried out, note: here 14. Includes
       one over root eight term */
    M[0]= (int)(coe[0]*ADD[9]);
    M[1]= (int)(coe[1]*ADD[14]);
    M[2]= (int)(coe[2]*ADD[11]);
    M[3]= (int)(coe[3]*ADD[4]);
    M[4]= (int)(coe[4]*ADD[15]);
    M[5]= (int)(coe[5]*ADD[5]);
    M[6]= (int)(coe[6]*ADD[12]);
```

Program 9.4a continued

```
    M[7]= (int)(coe[7]*ADD[16]);
    M[8]= (int)(coe[8]*ADD[13]);
    M[9]= (int)(coe[9]*ADD[6]);
    M[10]=(int)(coe[10]*ADD[17]);
    M[11]=(int)(coe[11]*ADD[7]);

    /* post multiplication, additions + subtractions */
    block[i]=(short)(ADD[18]>>3);                       /* y(0) */
    block[i+4]=(short)(ADD[19]>>3);                     /* y(4) */
    block[i+2]=(short)((M[0]+M[1])>>15);                /* y(2) */
    block[i+6]=(short)((M[1]-M[2])>>15);                /* y(6) */
    postadd1= M[6]+M[7];
    postadd2= M[7]+M[8];
    block[i+7]= (short)((M[3]+M[4]+postadd1)>>15);      /* y(7) */
    block[i+5]= (short)((M[4]+M[5]+postadd2)>>15);      /* y(5) */
    block[i+1]= (short)((M[9]+M[10]-postadd1)>>15);     /* y(1) */
    block[i+3]= (short)((postadd2-M[10]-M[11])>>15);    /* y(3) */
  }
}
```

```
/****************************************************/
/* idct.c performs a 8 point 2D Inverse DCT function */
/****************************************************/

/* externals */
#include "dct_main.h"
extern unsigned char image_in[IMAGE_SIZE];
extern unsigned char image_out[IMAGE_SIZE];
extern short block[BLOCK_SIZE];
extern const short coe[12];

void idct(void)
{
  short z[8],ADD[7];
  int PA[4];
  int M[13];
  int i,j;

  for(j=0;j<8;j++)                                      /* Rows */
  {
    /* pre-additions */
    ADD[0]=block[56+j]+block[40+j];                     /* y(7)+y(5) */
    ADD[1]=block[56+j]-block[8+j];                      /* y(7)-y(1) */
    ADD[2]=block[24+j]+block[40+j];                     /* y(3)+y(5) */
    ADD[3]=block[j]+block[32+j];                        /* y(0)+y(4) */
    ADD[4]=block[16+j]+block[48+j];                     /* y(2)+y(6) */
    ADD[5]=block[24+j]-block[8+j];                      /* y(3)-y(1) */
    ADD[6]=ADD[0]+ADD[5];

    /* multiplications */
    M[0]=(int)(coe[0] * block[16+j]);                   /* A*y(2) */
    M[1]=(int)(coe[1] * ADD[4]);
    M[2]=(int)(coe[2]*block[48+j]);                     /* C*y(6) */
```

(b) `idct.c`

Program 9.4b continued

```
    M[3]=(int)(coe[3]*block[56+j]);       /* D*y(7) */
    M[4]=(int)(coe[4]*ADD[0]);
    M[5]=(int)(coe[5]*block[40+j]);       /* F*y(5) */
    M[6]=(int)(coe[6]*ADD[1]);
    M[7]=(int)(coe[7]*ADD[6]);
    M[8]=(int)(coe[8]*ADD[2]);
    M[9]=(int)(coe[9]*block[8+j]);        /* J*y(1) */
    M[10]=(int)(coe[10]*ADD[5]);
    M[11]=(int)(coe[11]*block[24+j]);     /* L*y(3) */
    M[12]=(int)(block[32+j]<<1);          /* 2*y(4) */

    /* post additions */
    PA[0]=((M[0]+M[1])>>12);
    PA[1]=(int)ADD[3]-M[12];
    PA[2]=M[4]+M[7];
    PA[3]=M[7]-M[10];

    z[0]=ADD[3]+(short)PA[0];
    z[1]=ADD[3]-(short)PA[0];
    z[2]=(short)PA[1]+(short)((M[1]-M[2])>>12);
    z[3]=(short)PA[1]+(short)((M[2]-M[1])>>12);
    z[4]=(short)((PA[2]+M[6]+M[3])>>12);
    z[5]=(short)((PA[2]+M[5]+M[8])>>12);
    z[6]=(short)((PA[3]+M[6]+M[9])>>12);
    z[7]=(short)((PA[3]+M[8]-M[11])>>12);
    block[j]=z[0]+z[4];                   /* x(0) */
    block[8+j]=z[2]-z[5];                 /* x(1) */
    block[16+j]=z[3]+z[7];                /* x(2) */
    block[24+j]=z[1]+z[6];                /* x(3) */
    block[32+j]=z[1]-z[6];                /* x(4) */
    block[40+j]=z[3]-z[7];                /* x(5) */
    block[48+j]=z[2]+z[5];                /* x(6) */
    block[56+j]=z[0]-z[4];                /* x(7) */
  }
  for(i=0;i<64;i+=8)    /* Columns */
  {
    /* pre-additions */
    ADD[0]=block[i+7]+block[i+5];         /* y(7)+y(5) */
    ADD[1]=block[i+7]-block[i+1];         /* y(7)-y(1) */
    ADD[2]=block[i+3]+block[i+5];         /* y(3)+y(5) */
    ADD[3]=block[i]+block[i+4];           /* y(0)+y(4) */
    ADD[4]=block[i+2]+block[i+6];         /* y(2)+y(6) */
    ADD[5]=block[i+3]-block[i+1];         /* y(3)-y(1) */
    ADD[6]=ADD[0]+ADD[5];

    /* multiplications */
    M[0]=(int)(coe[0]*block[i+2]);        /* A*y(2) */
    M[1]=(int)(coe[1]*ADD[4]);
    M[2]=(int)(coe[2]*block[i+6]);        /* C*y(6) */
    M[3]=(int)(coe[3]*block[i+7]);        /* D*y(7) */
    M[4]=(int)(coe[4]*ADD[0]);
    M[5]=(int)(coe[5]*block[i+5]);        /* F*y(5) */
    M[6]=(int)(coe[6]*ADD[1]);
    M[7]=(int)(coe[7]*ADD[6]);
    M[8]=(int)(coe[8]*ADD[2]);
```

Program 9.4b continued

```
    M[9]=(int)(coe[9]*block[i+1]);        /* J*y(1) */
    M[10]=(int)(coe[10]*ADD[5]);
    M[11]=(int)(coe[11]*block[i+3]);      /* L*y(3) */
    M[12]=(int)(block[i+4]<<1);           /* 2*y(4) */

    /* post additions */
    PA[0]=(int)((M[0]+M[1])>>12);
    PA[1]=(int)((ADD[3]-M[12])>>3);
    PA[2]=M[4]+M[7];
    PA[3]=M[7]-M[10];

    z[0]=(ADD[3]+(short)PA[0])>>3;
    z[1]=(ADD[3]-(short)PA[0])>>3;
    z[2]=(short)PA[1]+(short)((M[1]-M[2])>>15);
    z[3]=(short)PA[1]+(short)((M[2]-M[1])>>15);
    z[4]=(short)((PA[2]+M[6]+M[3])>>15);
    z[5]=(short)((PA[2]+M[5]+M[8])>>15);
    z[6]=(short)((PA[3]+M[6]+M[9])>>15);
    z[7]=(short)((PA[3]+M[8]-M[11])>>15);

    block[i+0]=(z[0]+z[4]);               /* x[0] */
    block[i+1]=(z[2]-z[5]);               /* x[1] */
    block[i+2]=(z[3]+z[7]);               /* x[2] */
    block[i+3]=(z[1]+z[6]);               /* x[3] */
    block[i+4]=(z[1]-z[6]);               /* x[4] */
    block[i+5]=(z[3]-z[7]);               /* x[5] */
    block[i+6]=(z[2]+z[5]);               /* x[6] */
    block[i+7]=(z[0]-z[4]);               /* x[7] */
  }
}
```

```
/*************************************/
/* Constants used within the Program  */
/*************************************/

#ifndef DCT_MAIN_H

/*******************************************************/
/* dct_main performs a dct and idct function on an image. */
/*******************************************************/

/*************************************/
/* include files                      */
/*************************************/
#include <stdio.h>
#include <stdlib.h>
#include "dct_main.h"                     /* Includes and Constants used */
#pragma DATA_SECTION (image_in,"myvar0")
#pragma DATA_SECTION (image_out,"myvar1")
#include "scenary.h"                      /* an h file containing input image as a 1D array */
```

(c) dct_main.c

Program 9.4c continued

```
/* 1D array to hold output image */
unsigned char image_out[IMAGE_SIZE];

/* 1D array to hold the current block */
short block[BLOCK_SIZE];

/***************************************************/
/* Q12 DCT coefficients (actual coefficient x 2^12 ) */
/***************************************************/
const short coe[12]={3135,2217,7568,8410,-1598,6149,-10498,4816,-3686,-12586,8035,-1223};

/*************************************/
/* FUNCTIONS USED                    */
/*************************************/
/* dct.c */
void dct(void);

/* idct.c */
void idct(void);
/*************************************/
/* MAIN FUNCTION                     */
/*************************************/
void main()
{
  int i,x;

  /*************************************/
  *      FORWARD DCT/ INVERSE DCT      */
  *************************************/
  /* block by block processing */
  for(i=0;i<IMAGE_SIZE;i+=BLOCK_SIZE)
  {
    /* get the block from the input image */
    for(x=0;x<BLOCK_SIZE;++x)
      block[x] = (short) image_in[i+x];

    /* perform FDCT on this block */
    dct();

    /* perform IDCT on this block */
    idct();

    /* store block to output image */
    for(x=0;x<BLOCK_SIZE;++x)
    {
      if(block[x]<0)
        image_out[i+x]=(unsigned char) (-block[x]);
      else
        image_out[i+x]=(unsigned char) block[x];
     }
  }
}
/*************************************/
```

9.5 TMS320C6201 EVM–PC host communication

It has been shown that in order to display an image stored in the DSP memory, the image had to be first saved to a file using the debugger function, then displayed using MATLAB or other software. In this section the EVM–host communication is used to display the images as they are processed. Let us now study how the EVM and host communicate with each other.

The EVM and the host (PC) communicate through a PCI controller (AMCC S5933) also referred to in the AMCC literature as the MatchMaker (www.amcc.com). The AMCC S5933 communicates with the PC through the PCI local bus and communicates with the DSP through what is known as the Add-On bus as shown in Figure 9.5. The data transfer between the DSP and the host can be accomplished via the mailbox registers, FIFOs or the pass-through interface.

The pass-through interface is not described in this book and the reader is referred to the user guide SPRU269.

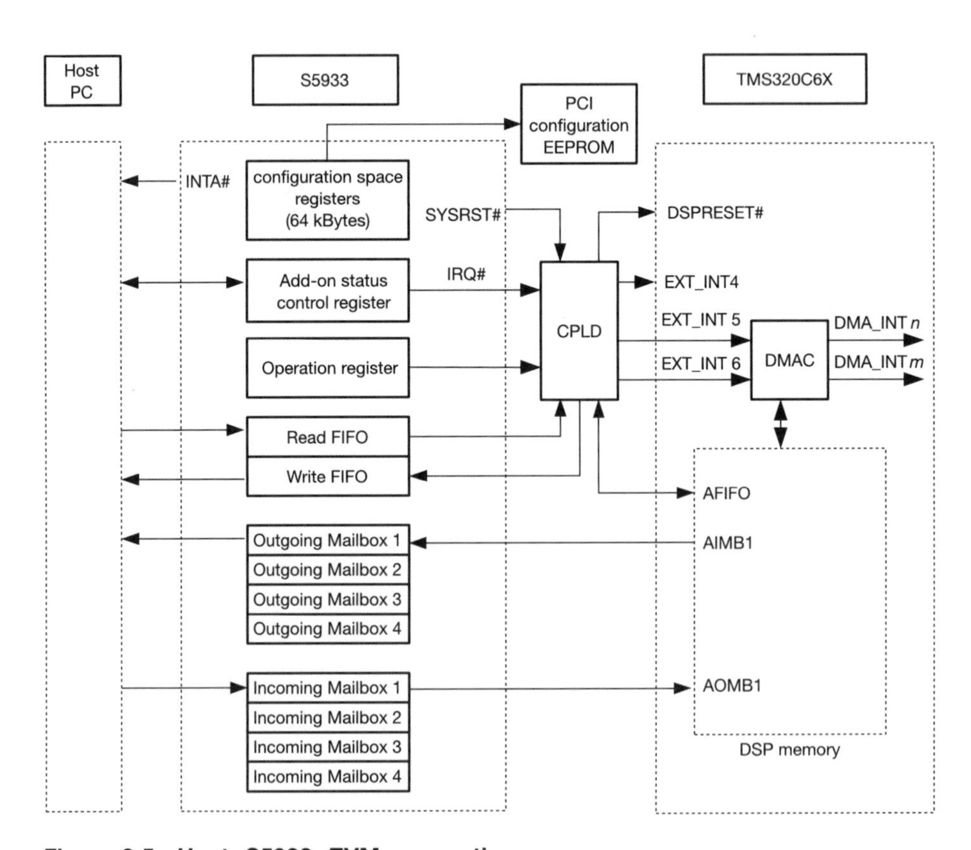

Figure 9.5 Host–S5933–EVM connections

9.5.1 Communication through the mailbox

There are two sets of four 32-bit mailbox registers for data transfer between the host and the PC. The Add-On Interrupt Status and Control Register (AINT) of the S5933 controls the functionality of these mailboxes (see Figure 9.6). Data sent from the host to the DSP can be conveyed through any one of the mailboxes known as the 'incoming' mailboxes, whereas data sent from the DSP to the host can be done through any one of the four mailboxes known as the 'outgoing' mailboxes. The two mailboxes used for communication (one incoming mailbox and one outgoing mailbox) are selected through the status and control register.

The AINT register is divided into two 16-bit sections:

(1) (16 LSBs: Interrupt Selection) This provides the source and method used to produce an interrupt.
(2) (16 MSBs: Interrupt Status) This is used for monitoring the source interrupts.

The AINT register is located in the address 0x01700038 (for MAP1) and is initialised to 0x00001010, which means that the incoming mailbox 1 is used (AINT[3:2] = 00). When the byte 0 becomes full, an interrupt is generated

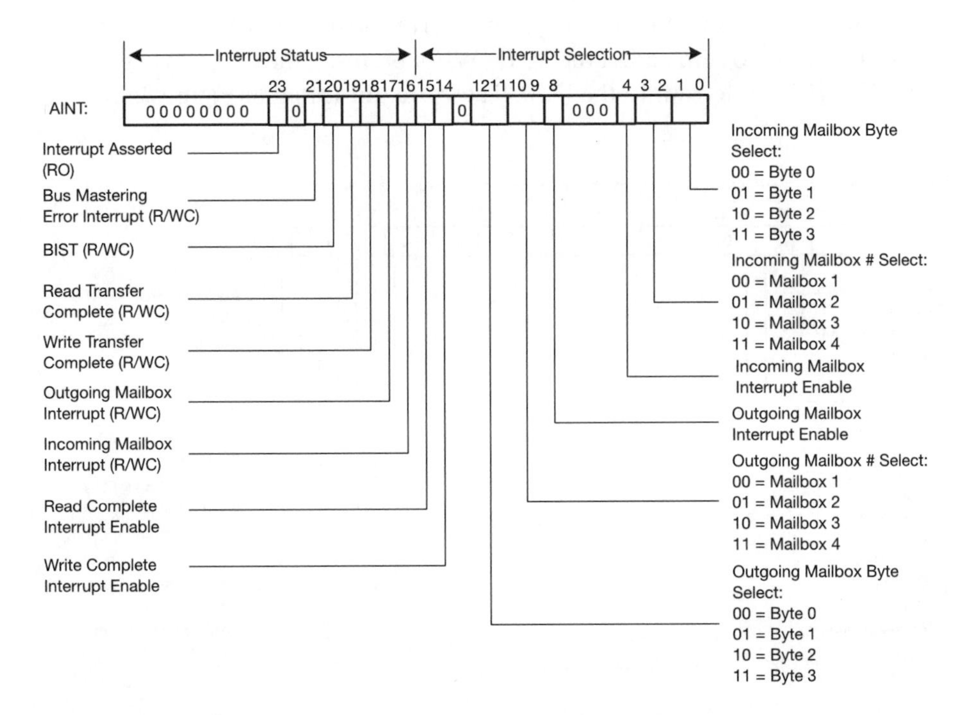

Figure 9.6 Add-On Interrupt Control/Status Register (AINT)

(AINT[1:0] = 00) if the interrupt is enabled (AINT[4] = 1). Bits AINT[8] to AINT[12] are for the outgoing mailbox counterparts.

In the event of the incoming mailbox (mailbox 1) becoming full or the outgoing mailbox (mailbox 1) becoming empty, the outgoing or incoming mailbox interrupt status bits (AINT[16], AINT[17]) are set to 1. An interrupt to the DSP and the host will be generated through the `IRQ#` and the `INTA#` respectively, as shown in Figure 9.6.

9.5.2 Communication through the FIFOs

There are two sets of eight 32-bit FIFOs. One set is used for data transfer from the host to the DSP and is known as the FIFO read, and the other is used for data transfer from the DSP to the host and is known as the FIFO write.

The Add-On General Control/Status Register (AGCSTS) controls the functionality of the FIFOs (see Figure 9.7). The DMA access to the AGCSTS register is through the CPLD memory-mapped FIFOSTAT register (located at address 0x1380018 for MAP0 and 0x1780018 for MAP1) as shown in Table 9.4.

When the data is written from the host to the FIFO **read**, the RDEMPTY flag in the AGCSTS will be set to zero to indicate that the FIFO **read** contains data and, if the PCIMREM bit is set, the CPLD will generate an EXT_INT5 to the DSP (see Figure 9.8). Similarly, if the FIFO **write** is not full, it will set the WRFULL flag to zero and, if the PCIMWEN bit is set, the CPLD will generate an EXT_INT6 interrupt to the DSP. These are illustrated in Figure 9.9.

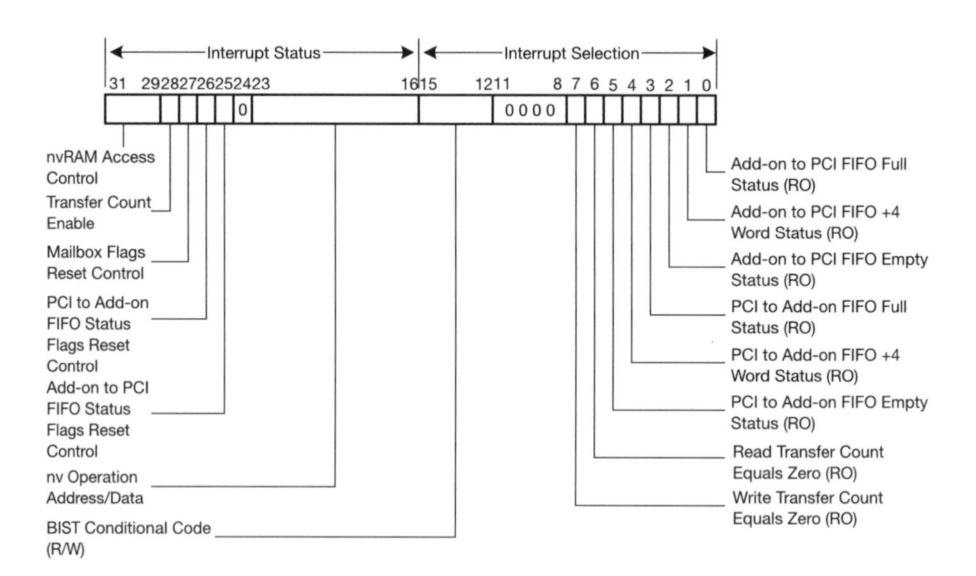

Figure 9.7 Add-On General Control/Status Register (AGCSTS)

Bit	Name (access)	Description
0	PCIMWEN (R/W)	PCI master write enable (0 = to disable, 1 = to enable)
1	PCIMREN (R/W)	PCI master read enable (0 = to disable, 1 = to enable)
2	WRFULL (R)	PCI controller write FIFO full (0 = not full, 1 = full)
3	RDEMPTY (R)	PCI controller read FIFO full (0 = not empty, 1 = empty)
4	PCIMWINT (R)	PCI master write interrupt (0 = not to activate, 1 = to activate)
5	PCIMRIT (R)	PCI master read interrupt (0 = not to activate, 1 = to activate)
6	–	–
7	–	–

Table 9.4 DSP FIFOSTAT register

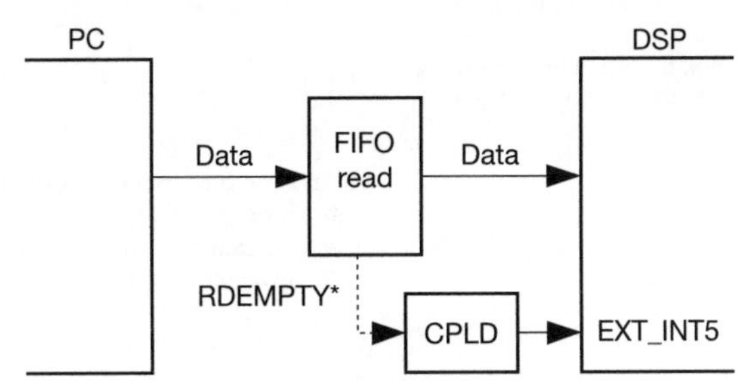

Figure 9.8 FIFO read interrupt mechanism

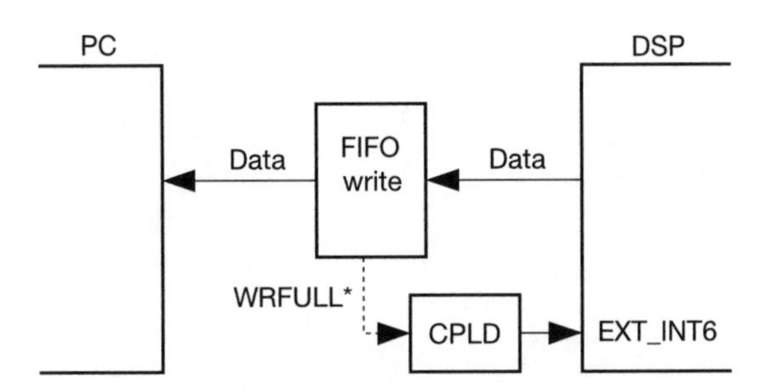

Figure 9.9 FIFO write interrupt mechanism

9.5.3 Demonstration program

A demonstration program, `host.exe`, demonstrates the implementation of the DCT, IDCT, quantisation, and host–DSP communication. There are three main programs:

(1) `host.c`: This program is compiled under Microsoft Visual C++ version 5.0 and its task is to perform the initialisations required on the PC side and call functions such as `DCT()`, `IDCT()` and `quant()` which are written in the `dsp.c` program.

DSP side	PC side
(0) The DSP waits The DSP waits for a mailbox command.	
	(1) The host sends a command message to the DSP: The host writes the required command into the mailbox by using the `evm6x_send_message()`. This will set the memory location 0x1700038 bit 16 to 1.
(2) The DSP decodes the command message The DSP detects the 0x1700038 bit 16 is set. When the mailbox message is present, the host reads the mailbox request. The DSP reads the mailbox to determine which operation to perform (DCT, IDCT, etc.) using the `GetHostReadReq()` function.	
	(3) The host prepares the data to be sent to the DSP through the FIFO The host sends 64 words to the DSP through the FIFO by using the `evm_write()` and then waits by using the `evm6x_retrieve_message()` function. When the host writes a word to the FIFO, an interrupt is generated (EXT_INT5).
(4) The DSP reads the data from the FIFO using the DMA The DMA keeps reading from the FIFO. When the DMA has transferred the 64 words an interrupt (DMA_INT2) is generated. The DSP polls until the DMA_INT2 flag is set and then clears it.	
(5) The DSP processes and prepares the data to be sent back to the host through the DMA. On completion, it sends a message to the host The `RcvFlag=1` is detected. The DSP transfers the data to be processed from `bufin[]` to `block[]`. The DSP processes the function required and copies the `block[]` processed to the `bufout[]`. The DSP prepares the DMA and starts it using the `sendDMA()` function (the DMA0 is used). The DSP sends a message to the host though the mailbox using the `RequestHostWrite()` function.	
	(6) The host receives the message The host receives the message from the DSP by using the function `evm6x_retrieve_message`.
	(7) The host reads the data The host reads the 64 words from the FIFO using the `evm6x_read()` function. **After all the 64 words are read, it generates an interrupt to the DMA_INT1.**
(8) The DSP polls until the DMA_INT1 flag is set Then this flag is cleared and the whole transaction is completed.	

Table 9.5 Brief description of the PC–DSP communication process

(2) dsp.c: This program is written for the 'C6201 processor and provides all initialisations required on the DSP side, routines for communication with the host and the functions such as DCT(), IDCT() and quant().
(3) scenary.h: This file contains the data for the image to be processed.

To aid understanding of how the host.c and dsp.c programs interact, see Table 9.5. The host.c and dsp.c files are shown in Program 9.5.

Table 9.5 is divided into two halves: one side (host.c) shows the main operations performed on the host and the other side shows the operations performed on the EVM. The numbers inside the parentheses represent the order by which the events occur. All the source and executable files can be found in F:\DSPCODE\DCT (Chap9)\DCTDMA\. When the DCT and IDCT processing is completed the host.exe will generate two displays, one representing the original image and the other the processed image (see Figure 9.10).

(a)

(b)

Figure 9.10 (a) Original image and (b) processed image

```
/*-------------------------------------------------------------------------------------
File:        dctdma.c
Description: DSP program to perform host request.
             Mainly processes 64 32-bit word block from host.
             Transfer between dsp and host via pci-dma.
             As DMA is done with PC, make sure the transfer blocks are situated outside the
             regularly accessed data region. In this case the dma blocks are in external
             memory.
-------------------------------------------------------------------------------------*/
```

(a) dctdma.c

Program 9.5 (a) dctdma.c; (b) dcthost.c

Program 9.5a continued

```
#include <common.h>
#include <cache.h>
#include <mcbsp.h>
#include <mcbspdrv.h>
#include <codec.h>
#include <dma.h>
#include <ctype.h>
#include <pci.h>
#include <time.h>
#include <intr.h>
#include <board.h>
#include <emif.h>
#include <regs.h>

#define BUFSIZE 1024
#define IMAGE_LEN 256
#define IMAGE_SIZE (IMAGE_LEN*IMAGE_LEN)
#define BLOCK_LEN 8
#define BLOCK_SIZE (BLOCK_LEN*BLOCK_LEN)

/**************************************************
                 MailBox Routines
**************************************************/
void MailBox_Setup (void)
{
  int Dummy;

  (*(volatile unsigned int *)0x01700038) = 0x00000010;  // Choose AIMB1, AOMB1
  Dummy = (*(volatile unsigned int *)0x01700000);
  (*(volatile unsigned int *)0x01700038) |= 0x00030000; // Clear previous mailbox status.
}

int MailBox_GetCommand (void)
{
  int HostCommand;

  while (!((*(volatile unsigned int *)0x01700038) & 0x00010000));
  HostCommand = (*(volatile unsigned int *)0x01700000);
  (*(volatile unsigned int *)0x01700038) |= 0x00010000;

  return HostCommand;
}

void MailBox_SendCommand (int HostCommand)
{
  (*(volatile unsigned int *)0x01700010) = (unsigned int) HostCommand;
  //===== Optional, we can wait until the host read the message. =====
  // while (!((*(volatile unsigned int *)0x01700038) & 0x00020000));
  // (*(volatile unsigned int *)0x01700038) |= 0x00020000;
}

/**************************************************
                  DMA Routines
**************************************************/
#pragma DATA_SECTION (ReceiveDmaBuffer, ".edata")
#pragma DATA_SECTION (SendDmaBuffer, ".edata")
```

Program 9.5a continued

```
int ReceiveDmaBuffer[64];
int SendDmaBuffer[64];

void DMA_Setup (void)
{
  DMA_STOP (2); DMA_STOP (0);
  *(volatile unsigned int *)(cpldBaseAddr+CPLD_FIFOSTAT_OFFSET) &= 0xfc;
  //--- Send DMA (Channel 0) ---
  DMA0_PRIMARY_CTRL =
  (DMA_RELOAD_NONE << 30)      // Don't reload destination address, fixed
    | (DMA_RELOAD_NONE << 28)  // Don't reload source address, dsp will reload.
    | (0               << 27)  // Emode=0, DMA not halted during emulation halt.
    | (0               << 26)  // FS=0, No frame sync.
    | (1               << 25)  // TCINT=1, enable dma interrupt.
    | (DMA_DMA_PRI     << 24)  // DMA has priority of CPU on internal bus.
    | (SEN_EXT_INT6    << 19)  // Write Sync from external interrupt pin 6 (evm6x requirement).
    | (SEN_NONE        << 14)  // No Read Sync.
    | (DMA_GNDX_A      << 13)  // Use global index register A as programmable index.
    | (0               << 12)  // no count reload.
    | (DMA_SPLIT_DIS   << 10)  // Disable split channel mode.
    | (DMA_ESIZE32     <<  8)  // Set element size to 32-bit.
    | (DMA_ADDR_NO_MOD <<  6)  // Fix destination address.
    | (DMA_ADDR_INC    <<  4)  // Auto-increment source address after each transfer.
    | (DMA_STOP_VAL         ); // Stop dma now. Start later, after configuration.

  DMA0_SECONDARY_CTRL =
      (1  << 15)               // Clear any prior WSYNC status.
    | (1  << 13)               // Clear any prior RSYNC status.
    | (1  <<  7);              // Enable block-transfer-end interrupt.
  DMA0_DEST_ADDR = PCI_MAP1_FIFO_ADDR;  // Global address set after evm_init() is called.
  intr_map (CPU_INT10, ISN_DMA_INT0);
  INTR_CLR_FLAG (CPU_INT10);

  //--- Receive DMA (Channel 2) ---
  DMA2_PRIMARY_CTRL =
  (DMA_RELOAD_NONE    << 30)    // Don't reload destination address, fixed.
    | (DMA_RELOAD_NONE << 28)   // Don't reload source address, dsp will reload.
    | (0               << 27)   // Emode=0, DMA not halted during emulation halt.
    | (0               << 26)   // FS=0, No frame sync.
    | (1               << 25)   // TCINT=1, enable dma interrupt.
    | (DMA_DMA_PRI     << 24)   // DMA has priority of CPU on internal bus.
    | (SEN_NONE        << 19)   // No Write Sync.
    | (SEN_EXT_INT5    << 14)   // Read Sync from external interrupt pin 5 (EVM6X requirement).
    | (DMA_GNDX_A      << 13)   // Use global index register A as programmable index.
    | (0               << 12)   // no count reload.
    | (DMA_SPLIT_DIS   << 10)   // Disable split channel mode.
    | (DMA_ESIZE32     <<  8)   // Set element size to 32-bit.
    | (DMA_ADDR_INC    <<  6)   // Auto-increment destination address after each transfer.
    | (DMA_ADDR_NO_MOD <<  4)   // Fix source address.
    | (DMA_STOP_VAL         );  // Stop dma now. Start later, after configuration.

  DMA2_SECONDARY_CTRL =
      (1               << 15)   // Clear any prior WSYNC status.
    | (1               << 13)   // Clear any prior RSYNC status.
    | (1               <<  7);  // Enable block-transfer-end interrupt.
```

Program 9.5a continued

```
  DMA2_SRC_ADDR = PCI_MAP1_FIFO_ADDR;  // Global address set after evm_init() is called.
  intr_map (CPU_INT11, ISN_DMA_INT2);
  INTR_CLR_FLAG (CPU_INT11);

  return;
}

void DMA_Receive (int *block)
{
  int i;

  DMA2_DEST_ADDR = (unsigned int) ReceiveDmaBuffer;
  DMA2_XFER_COUNTER = 64;
  DMA_START (2);
  *(volatile unsigned int *)(cpldBaseAddr+CPLD_FIFOSTAT_OFFSET) |= 0x02;
  while (!INTR_CHECK_FLAG(CPU_INT11));
  DMA2_SECONDARY_CTRL &= 0xffffffbf;
  INTR_CLR_FLAG (CPU_INT11);
  *(volatile unsigned int *)(cpldBaseAddr+CPLD_FIFOSTAT_OFFSET) &= 0xfd;
  memcpy (block, ReceiveDmaBuffer, 64 * sizeof (int));
  DMA_STOP (2);
}

void DMA_Send (int *block)
{
  int i;

  memcpy (SendDmaBuffer, block, 64 * sizeof (int));
  DMA0_SRC_ADDR = (unsigned int) SendDmaBuffer;
  DMA0_XFER_COUNTER = 64;
  DMA_START (0);
  *(volatile unsigned int *)(cpldBaseAddr+CPLD_FIFOSTAT_OFFSET) |= 0x01;
  while (!INTR_CHECK_FLAG(CPU_INT10));
  INTR_CLR_FLAG (CPU_INT10);
  DMA0_SECONDARY_CTRL &= 0xffffffbf;
  *(volatile unsigned int *)(cpldBaseAddr+CPLD_FIFOSTAT_OFFSET) &= 0xfe;
  DMA_STOP (0);
}


/**************************************************
             DCT and IDCT functions
**************************************************/
/* Coefficients in Q1.12 format */
const int coe[8][8]=
{
  1448,  1448,  1448,  1448,  1448,  1448,  1448,  1448,
  2008,  1702,  1137,   399,  -399, -1137, -1702, -2008,
  1892,   783,  -783, -1892, -1892,  -783,   783,  1892,
  1702,  -399, -2008, -1137,  1137,  2008,   399, -1702,
  1448, -1448, -1448,  1448,  1448, -1448, -1448,  1448,
  1137, -2008,   399,  1702, -1702,  -399,  2008, -1137,
   783, -1892,  1892,  -783,  -783,  1892, -1892,   783,
   399, -1137,  1702, -2008,  2008, -1702,  1137,  -399,
};
```

Program 9.5a continued

```
/* Quantiser in integer format. */
const int quant[64] =
{
  16,  11,  10,  16,  24,  40,  51,  61,
  12,  12,  14,  19,  26,  58,  60,  55,
  14,  13,  16,  24,  40,  57,  69,  56,
  14,  17,  22,  29,  51,  87,  80,  62,
  18,  22,  37,  56,  68, 109, 103,  77,
  24,  35,  55,  64,  81, 104, 113,  92,
  49,  64,  78,  87, 103, 121, 120, 101,
  72,  92,  95,  98, 112, 100, 103,  99
};

/* Reciprocal of quantiser in Q1.15 format. */
const int rquant15 [64] =
{
  0x0800, 0x0ba2, 0x0ccc, 0x0800, 0x0555, 0x0333, 0x0282, 0x0219,
  0x0aaa, 0x0aaa, 0x0924, 0x06bc, 0x04ec, 0x0234, 0x0222, 0x0253,
  0x0924, 0x09d8, 0x0800, 0x0555, 0x0333, 0x023e, 0x01da, 0x0249,
  0x0924, 0x0787, 0x05d1, 0x0469, 0x0282, 0x0178, 0x0199, 0x0210,
  0x071c, 0x05d1, 0x0375, 0x0249, 0x01e1, 0x012c, 0x013e, 0x01a9,
  0x0555, 0x03a8, 0x0253, 0x0200, 0x0194, 0x013b, 0x0121, 0x0164,
  0x029c, 0x0200, 0x01a4, 0x0178, 0x013e, 0x010e, 0x0111, 0x0144,
  0x01c7, 0x0164, 0x0158, 0x014e, 0x0124, 0x0147, 0x013e, 0x014a
};

#pragma CODE_SECTION (dct,".iprog")
far void dct(int *block)
{
  int i,j,x,y;
  int value[8];

  /* perform 1D DCT on the rows */
  for(i=0;i<64;i+=8)
  {
    for(y=0;y<8;++y)
    {
      value[y] = 0;
      for(x=0;x<8;++x)
        value[y] += (coe[y][x]*block[i+x]);
    }
    for(y=0;y<8;++y)
      block[i+y] = (value[y]>>12);
  }
  /* perform 1D DCT on the columns */
  for(j=0;j<8;j++)
  {
    for(y=0;y<8;++y)
    {
      value[y]=0;
      for(x=0;x<8;++x)
        value[y] += (coe[y][x]*block[j+(x*8)]);
    }
```

Program 9.5a continued

```
    for(y=0;y<8;++y)
      block[j+(y*8)] = (value[y]>>12);
  }
}

#pragma CODE_SECTION (idct,".iprog")
far void idct(int *block)
{
  int i,j,x,y;
  int value[8];

  /* perform 1D IDCT on the rows */
  for(i=0;i<64;i+=8)
  {
    for(y=0;y<8;++y)
    {
      value[y] = 0;
      for(x=0;x<8;++x)
        value[y] += (int)(coe[x][y]*block[i+x]);
    }
    for(y=0;y<8;++y)
      block[i+y] = (short)(value[y]>>12);
  }
  /* perform 1D IDCT on the columns */
  for(j=0;j<8;j++)
  {
    for(y=0;y<8;++y)
    {
      value[y] = 0;
      for(x=0;x<8;++x)
        value[y] += (int)(coe[x][y]*block[j+(x*8)]);
    }
    for(y=0;y<8;++y)
      block[j+(y*8)] = (short)(value[y]>>12);
  }

  for(i=0;i<64;i++)
  {
    if (block[i] < 0) block[i] = 0;
    if (block[i] > 255) block[i] = 255;
  }
}

#pragma CODE_SECTION (ExecuteCommand,".iprog")
far void ExecuteCommand (unsigned int msg, int *block)
{
  int i;

  switch (msg) /* Perform appropriate operation as required by host. */
  {
    case 0:
      for (i=0; i<64; i++)
        block[i]=block[i]^0xff;
      break;
```

Program 9.5a continued

```
    case 1:                     // Perform DCT.
      dct (block);
      break;
    case 2:                     // Perform IDCT
      idct (block);
      break;
    case 3:                     // Perform DCT with JPEG quantisation.
      dct (block);
      for (i=0; i<64; i++)
      {
        /* Resulting values have 15 decimal places.             */
        /* Hence shift 15 spaces left to obtain integral value.  */
        /* For negative values, 0.5 is added to the integral     */
        /* so that values within -1 and 0 can be round down to 0 */
        /* instead of -1.                                        */
        block[i] = ((short)block[i]*(short)rquant15[i])>>14;
        if (block[i]<0) block[i] +=1;
        block[i] = block[i] >> 1;
      }
      break;
    case 4:   // Perform IDCT with JPEG dequantisation.
      for (i=0; i<64; i++)
        block[i]=block[i]*quant[i];
      idct (block);
      break;
    default: /* does nothing to the block */
      break;
  }
}

/***************************************************
                   Main Loop
***************************************************/
void main(int argc, char *argv[])
{
  int msg;
  int block [64];

  evm_init();

  DMA_Setup ();
  MailBox_Setup ();

  While (1)
  {
    //--- Wait for Host message and read the message ---
    msg = MailBox_GetCommand ();

    //--- Start DMA to read 64-word block into input buffer ---
    DMA_Receive (block);

    //--- Process the 64-word block according to the host command.
    ExecuteCommand (msg, block);

    //--- Inform host of 64-word block ready ---
    MailBox_SendCommand (64);
```

Program 9.5a continued

```
    //--- Start DMA to write the processed 64-word block to host ---
    DMA_Send (block);
  }
  return;
}
```

```
/*------------------------------------------------------------------------------------
  File:        dcthost.c
  Description: A p.c. program to work with dctdma.c
               Perform a DCT IDCT on a picture using DSP.
               Blocks are sent and received in 64-word (32-bit) blocks via PCI-DMA.
               It loads the dsp program in DCTDMA.OUT before running.
               Note requires Microsoft(TM) Visual C++ 5.0 to run.
               Also requires to link with evm6x.lib.
               Make sure evm6x.dll is available.
               If there is error reading, increase the timeout value in
               the evm6x_set_timeout (time_out_ms) function.
------------------------------------------------------------------------------------*/

#include <windows.h>
#include <stdio.h>
#include <conio.h>
#include <process.h>
#include "scenary.h"
#include "evm6xdll.h"

/*******************************************************
                 Display Routines.
*******************************************************/
HWND hwnd1, hwnd2;
HBITMAP hbitmap1, hbitmap2;
HDC memdc1, memdc2;
volatile BOOL bInit;

LRESULT CALLBACK DisplayWndProc (HWND hwnd, UINT uMsg, WPARAM wParam, LPARAM lParam)
{
  PAINTSTRUCT ps;
  HDC hdc;

  Switch (uMsg)
  {
    case WM_PAINT:
      BeginPaint (hwnd, &ps);
      hdc = GetDC (hwnd);
      BitBlt (hdc, 0, 0, 256, 256, (hwnd==hwnd1)?memdc1:memdc2, 0, 0, SRCCOPY);
      ReleaseDC (hwnd, hdc);
      EndPaint (hwnd,  &ps);
      break;
    default:
      return DefWindowProc (hwnd, uMsg, wParam, lParam);
  }
```

(b) dcthost.c

Program 9.5b continued

```
  return 0;
}

DWORD WINAPI DisplayThread (void *dummy)
{
  MSG msg;
  WNDCLASS wndClass;
  Char lpClassName[] = "DISPLAY_TEST_DCT";
  RECT rect;
  HDC hdc;

  Memset(&wndClass, 0, sizeof(wndClass));
  wndClass.style = CS_HREDRAW | CS_VREDRAW;
  wndClass.hInstance = 0;
  wndClass.hIcon = NULL;
  wndClass.hCursor = LoadCursor (NULL, IDC_ARROW);
  wndClass.hbrBackground = GetStockObject (BLACK_BRUSH);
  wndClass.lpszMenuName = NULL;
  wndClass.lpfnWndProc = DisplayWndProc;
  wndClass.lpszClassName = lpClassName;
  RegisterClass(&wndClass);

  Rect.top = rect.left = 0; rect.right = rect.bottom = 256;
  AdjustWindowRect (&rect, WS_CAPTION, FALSE);

  Hwnd1 = CreateWindow (lpClassName, "Raw Image Display",
    WS_POPUP|WS_CAPTION|WS_VISIBLE,
    0, 0,
    rect.right-rect.left, rect.bottom-rect.top,
    NULL, NULL, NULL, NULL);

  Hwnd2 = CreateWindow (lpClassName, "Reconstructed Image Display",
    WS_POPUP|WS_CAPTION|WS_VISIBLE,
    rect.right-rect.left, 0, rect.right-rect.left, rect.bottom-rect.top,
    NULL, NULL, NULL, NULL);

  Hdc = GetDC (hwnd1);
  memdc1 = CreateCompatibleDC (hdc);
  hbitmap1 = CreateCompatibleBitmap (hdc, 256, 256);
  SelectObject (memdc1, hbitmap1);
  SelectObject (memdc1, GetStockObject (BLACK_BRUSH));
  SelectObject (memdc1, GetStockObject (BLACK_PEN));
  Rectangle (memdc1, 0, 0, 256, 256);
  ReleaseDC (hwnd1, hdc);

  Hdc = GetDC (hwnd2);
  memdc2 = CreateCompatibleDC (hdc);
  hbitmap2 = CreateCompatibleBitmap (hdc, 256, 256);
  SelectObject (memdc2, hbitmap2);
  SelectObject (memdc2, GetStockObject (WHITE_BRUSH));
  SelectObject (memdc2, GetStockObject (WHITE_PEN));
  Rectangle (memdc2, 0, 0, 256, 256);
  ReleaseDC (hwnd2, hdc);
  BInit = TRUE;

  while (GetMessage(&msg, NULL, 0, 0))
```

Program 9.5b continued

```
  {
    TranslateMessage (&msg);
    DispatchMessage(&msg);
  }
  return msg.wParam;
}

void DisplayInit (void)
{
  DWORD dwId;
  BInit = FALSE;
  CreateThread (NULL, 0, DisplayThread, NULL, 0, &dwId);
  While (!bInit);
  return;
}

void DisplayBlock (int display, int row, int col, int *pRawBlock)
{
  int x, y;
  RECT rect;
  int int_pixel;
  unsigned char char_pixel;
  HWND hwnd;
  HDC memdc, hdc;
  int pBlock [64];
  int *ptr;

  memcpy (pBlock, pRawBlock, 64*sizeof (int));
  ptr = pBlock;
  if (display == 1)
  {
    hwnd = hwnd1;
    memdc = memdc1;
  }
  else
  {
    hwnd = hwnd2;
    memdc = memdc2;
  }

  for (y=0;y<8;y++)
  {
    for (x=0;x<8;x++)
    {
      int_pixel = *ptr++;
      if (int_pixel<0) char_pixel = 0;
      else if (int_pixel>255) char_pixel = 255;
      else char_pixel = (char) int_pixel;
      SetPixel (memdc, col*8+x, row*8+y, RGB (char_pixel, char_pixel, char_pixel));
    }
  }
  rect.left = col*8; rect.right=rect.left+7;
  rect.top = row*8; rect.bottom=rect.top+7;
  hdc = GetDC (hwnd);
  SendMessage (hwnd, WM_PAINT, (WPARAM) hdc, 0);
```

Program 9.5b continued

```
  ReleaseDC (hwnd, hdc);

  return;
}

/*********************************************************
                    DSP Routines.
*********************************************************/
#define LOAD_DSP_CODE
HANDLE hBoard;
HANDLE hEvent;
#define DSP_OTHERS -1
#define DSP_DIRECT 0x00
#define DSP_DCT 0x01
#define DSP_IDCT 0x02
#define DSP_DCT_QUANTIZE 0x03
#define DSP_IDCT_UNQUANTIZE 0x04

BOOL DspBoardInit (void)
{
  int dummy;

  #ifdef LOAD_DSP_CODE
  evm6x_reset_dsp (hBoard, HPI_BOOT);
  evm6x_init_emif (hBoard, NULL);
  if (!evm6x_coff_load (hBoard, NULL, "DCTDMA.OUT", FALSE, FALSE, FALSE))
  {
    evm6x_unreset_dsp (hBoard);
    return FALSE;
  }
  evm6x_unreset_dsp (hBoard);
  #endif
  evm6x_retrieve_message (hBoard, &dummy); // Clear previous mail-box message.

  return TRUE;
}

BOOL DspDriverInit (void)
{
  char str[80];

  hBoard = evm6x_open (0, FALSE);
  if (hBoard == INVALID_HANDLE_VALUE)
    return FALSE;

  evm6x_set_timeout (hBoard, 1000);
  evm6x_abort_write (hBoard);
  evm6x_clear_message_event (hBoard);

  sprintf (str, "%s%d", EVM6X_GLOBAL_MESSAGE_EVENT_BASE_NAME, 0);
  hEvent = OpenEvent (SYNCHRONIZE, FALSE, str);
  if (hEvent==NULL)
  {
    evm6x_close (hBoard);
    return FALSE;
  }
```

Program 9.5b continued

```
  return TRUE;
}

void DspDriverClose (void)
{
  CloseHandle (hEvent);
  evm6x_close (hBoard);
  return;
}

int DspCommand (int *input, int *output, int command)
{
  EVM6XDLL_MESSAGE msg;
  int count;
  int wait_ms = 3000; // Time out for dsp to process a block.

  //===== Send command to dsp via mailbox =====
  msg = command; if (!evm6x_send_message (hBoard, &msg)) return 1;

  //===== Send 64-word block via PCI-DMA to DSP for processing =====
  count = 256; if (!evm6x_write (hBoard, (unsigned long *) input, &count)) return 2;
  if (count!=256) return 3;

  //===== Wait for mailbox message from DSP =====
  while (!WaitForSingleObject (hEvent, 1))
  {
    wait_ms-=1;
    if (wait_ms == 0) return 4;
  }
  evm6x_retrieve_message (hBoard, &msg);

  //===== Read processed 64-word block via PCI-DMA from DSP =====
  count = 256; if (!evm6x_read (hBoard, (unsigned long *) output, &count)) return 5;
  if (count!=256) return 6;
  return 0;
}

/*********************************************************
                  Main Routines.
*********************************************************/
int main (int argc, char *argv[])
{
  int row, col;
  int x, y;
  int pRawBlock[64];
  int pDctBlock[64];
  int pReconBlock[64];
  int *pTmp;
  int nReturn;

  //===== Initialise the DSP driver, DSP board and display =====
  if (!DspDriverInit ())
  {
    printf ("Error Initialising Evm6x driver!\n");
    return -1;
  }
```

Program 9.5b continued

```
if (!DspBoardInit ())
{
  printf ("Error Initialising Evm6x board!\n");
  DspDriverClose ();
  return -1;
}
DisplayInit ();

//===== Perform DSP communications (each transaction involves 64 32-bit words =====
row = 0; col = 0;
while (1)
{
  /* User quits before program finishes */
  if (kbhit())
  {
    getch();
    break;
  }

  /* Collect 64-pixel block and arrange into contiguous block and display it. */
  pTmp = pRawBlock;
  for (y=0;y<8;y++)
    for (x=0;x<8;x++)
      *pTmp++ = (int) image_in[(col*8+x)+(row*8+y)*256];
    DisplayBlock (1, row, col, pRawBlock);

    /* Send a DSP command: DCT */
  if ((nReturn=DspCommand (pRawBlock, pDctBlock, DSP_DCT))!=0)
  {
    switch (nReturn)
    {
      case 1: printf ("Error in Mailbox Message (row %d, col %d)!\n", row, col); break;
      case 2: printf ("Error in DmaWrite (row %d, col %d)!\n", row, col); break;
      case 3: printf ("Cannot complete DmaWrite (row %d, col %d)!\n", row, col); break;
      case 4: printf ("DSP process takes too long (row %d, col %d)!\n", row, col); break;
      case 5: printf ("Error in DmaRead (row %d, col %d)!\n", row, col); break;
      case 6: printf ("Cannot complete DmaRead (row %d, col %d)!\n", row, col); break;
    }
    break;
  }

  /* Send a DSP command: IDCT */
  if ((nReturn=DspCommand (pDctBlock, pReconBlock, DSP_IDCT))!=0)
  {
    switch (nReturn)
    {
      case 1: printf ("Error in Mailbox Message (row %d, col %d)!\n", row, col); break;
      case 2: printf ("Error in DmaWrite (row %d, col %d)!\n", row, col); break;
      case 3: printf ("Cannot complete DmaWrite (row %d, col %d)!\n", row, col); break;
      case 4: printf ("DSP process takes too long (row %d, col %d)!\n", row, col); break;
      case 5: printf ("Error in DmaRead (row %d, col %d)!\n", row, col); break;
      case 6: printf ("Cannot complete DmaRead (row %d, col %d)!\n", row, col); break;
    }
    break;
  }
```

Program 9.5b continued

```
    /* Display processed block */
    DisplayBlock (2, row, col, pReconBlock);

    /* Next block */
    if (++col == 32)
    {
      col = 0;
      if (++row == 32)
      {
        DspDriverClose ();
        printf ("Done!. Press any key to quit\n");
        getch ();
        return 0;
      }
    }
  }
  //===== Close the connection with the DSP driver =====
  DspDriverClose ();
  return 0;
}
```

Appendix A

Optimisation of 1-D DCT and IDCT

A.1 1-D DCT (DCT)

Consider an 8×1 column vector with elements $x(0)$, $x(1)$, $x(2)$, $x(3)$, $x(4)$, $x(5)$, $x(6)$ and $x(7)$. We can derive a corresponding discrete cosine transform with eight coefficients $X(0)$, $X(1)$, $X(2)$, $X(3)$, $X(4)$, $X(5)$, $X(6)$ and $X(7)$ using the equation

$$X(k) = \frac{1}{2} C(k) \sum_{i=0}^{7} x(i) \cos\left[\frac{(2i+1)k\pi}{16}\right], \quad k = 0, 1, \ldots, 7$$

where

$$C(k) = \begin{cases} \dfrac{1}{\sqrt{2}} & \text{if } k = 0 \\ 1 & \text{otherwise} \end{cases}$$

We shall first replace $C(k)$ by $C'(k)$ as follows (this will simplify $X(0)$ and $X(4)$, as will be shown later):

$$X(k) = \frac{1}{2\sqrt{2}} C'(k) \sum_{i=0}^{7} x(i) \cos\left[\frac{(2i+1)k\pi}{16}\right], \quad k = 0, 1, \ldots, 7$$

where

$$C'(k) = \begin{cases} 1 & \text{if } k = 0 \\ \sqrt{2} & \text{otherwise} \end{cases}$$

As all terms contain $1/2\sqrt{2}$, we shall instead consider $X'(k)$ where:

$$X(k) = \frac{1}{2\sqrt{2}} X'(k)$$

and

$$X'(k) = \sum_{i=0}^{7} x(i) C'(k) \cos\left[\frac{(2i+1)k\pi}{16}\right], \quad k = 0, 1, \ldots, 7$$

Before optimisation notice that when $k = 0$, all the eight cosine terms are evaluated to be $\cos(0) = 1$ and since $C'(0) = 1$, the summation is simply

$$X'(0) = \sum_{i=0}^{7} x(i) = x(0) + x(1) + x(2) + x(3) + x(4) + x(5) + x(6) + x(7)$$

It is also worth noting that at $k = 4$, the cosine terms consist of odd multiples of $\pi/4$, which are either $1/\sqrt{2}$ or $-1/\sqrt{2}$. Since $C'(4)$ is $\sqrt{2}$, the coefficients of $x(i)$ are either 1 or -1:

$$\begin{aligned} X'(4) &= \sum_{i=0}^{7} x(i)\sqrt{2}\cos\left[(2i+1)\frac{\pi}{4}\right] \\ &= x(0) - x(1) - x(2) + x(3) + x(4) - x(5) - x(6) + x(7) \end{aligned}$$

So $X(0)$ and $X(4)$ can be evaluated without multiplications, just additions and subtractions.

We next compute the terms $C'(k)\cos[(2i+1)\pi k/16]$ for all values of i and k. Table A.1 shows that only six values are used (ignore signs and the 1 and -1); for simplicity we call them:

$$\begin{aligned} a &= \sqrt{2}\cos(\pi/16) \approx 1.387 \\ b &= \sqrt{2}\cos(2\pi/16) \approx 1.307 \\ c &= \sqrt{2}\cos(3\pi/16) \approx 1.176 \\ d &= \sqrt{2}\cos(6\pi/16) \approx 0.541 \\ e &= \sqrt{2}\cos(7\pi/16) \approx 0.276 \\ f &= \sqrt{2}\cos(5\pi/16) \approx 0.786 \end{aligned}$$

Now Table A.1 can be rewritten as shown in Table A.2.

Note that the rows with $k = 2$ and $k = 6$ in Table A.2 have only two factors (b and d, without considering the sign), whereas $k = 1, 3, 5, 7$ have four common 'factors', a, c, e, f. By regrouping them as shown in Table A.3 we can express the whole transform in matrix form:

i/k	0	1	2	3	4	5	6	7
0	1	1	1	1	1	1	1	1
1	1.387	1.176	0.786	0.276	−0.276	−0.786	−1.176	−1.387
2	1.307	0.541	−0.541	−1.307	−1.307	−0.541	0.541	1.307
3	1.176	−0.276	−1.387	−0.786	0.786	1.387	0.276	−1.176
4	1	−1	−1	1	1	−1	−1	1
5	0.786	−1.387	0.276	1.176	−1.176	−0.276	1.387	−0.786
6	0.541	−1.307	1.307	−0.541	−0.541	1.307	−1.307	0.541
7	0.276	−0.786	1.176	−1.387	1.387	−1.176	0.786	−0.276

Table A.1 DCT coefficients

i/k	0	1	2	3	4	5	6	7
0	1	1	1	1	1	1	1	1
1	*a*	*c*	*f*	*e*	−*e*	−*f*	−*c*	−*a*
2	*b*	*d*	−*d*	−*b*	−*b*	−*d*	*d*	*b*
3	*c*	−*e*	−*a*	−*f*	*f*	*a*	*e*	−*c*
4	1	−1	−1	1	1	−1	−1	1
5	*f*	−*a*	*e*	*c*	−*c*	−*e*	*a*	−*f*
6	*d*	−*b*	*b*	−*d*	−*d*	*b*	−*b*	*d*
7	*e*	−*f*	*c*	−*a*	*a*	−*c*	*f*	−*e*

Table A.2 DCT coefficients

i/k	0	1	2	3	4	5	6	7
0	1	1	1	1	1	1	1	1
4	1	−1	−1	1	1	−1	−1	1
2	*b*	*d*	−*d*	−*b*	−*b*	−*d*	*d*	*b*
6	*d*	−*b*	*b*	−*d*	−*d*	*b*	−*b*	*d*
7	*e*	−*f*	*c*	−*a*	*a*	−*c*	*f*	−*e*
5	*f*	−*a*	*e*	*c*	−*c*	−*e*	*a*	−*f*
1	*a*	*c*	*f*	*e*	−*e*	−*f*	−*c*	−*a*
3	*c*	−*e*	−*a*	−*f*	*f*	*a*	*e*	−*c*

Table A.3 DCT coefficients reordered

$$\begin{bmatrix} X'(0) \\ X'(4) \\ X'(2) \\ X'(6) \\ X'(7) \\ X'(5) \\ X'(1) \\ X'(3) \end{bmatrix} = \begin{bmatrix} 1 & 1 & 1 & 1 & 1 & 1 & 1 & 1 \\ 1 & -1 & -1 & 1 & 1 & -1 & -1 & 1 \\ b & d & -d & -b & -b & -d & d & b \\ d & -b & b & -d & -d & b & -b & d \\ e & -f & c & -a & a & -c & f & -e \\ f & -a & e & c & -c & -e & a & -f \\ a & c & f & e & -e & -f & -c & -a \\ c & -e & -a & -f & f & a & e & -c \end{bmatrix} \begin{bmatrix} x(0) \\ x(1) \\ x(2) \\ x(3) \\ x(4) \\ x(5) \\ x(6) \\ x(7) \end{bmatrix}$$

Notice that the 8×8 matrix is anti-symmetric about the line drawn vertically across the middle of the matrix. By factorising and making use of this symmetry, we can arrive at the following matrix that aims at minimising the number of multiplications (McGovern *et al.*, 1994):

$$\begin{bmatrix} X'(0) \\ X'(4) \\ X'(2) \\ X'(6) \\ X'(7) \\ X'(5) \\ -X'(1) \\ X'(3) \end{bmatrix} = \begin{bmatrix} 1 & 0 & 0 & 0 & 0 & 0 & 0 & 0 \\ 0 & 1 & 0 & 0 & 0 & 0 & 0 & 0 \\ 0 & 0 & b & d & 0 & 0 & 0 & 0 \\ 0 & 0 & d & -b & 0 & 0 & 0 & 0 \\ 0 & 0 & 0 & 0 & e & f & -a & c \\ 0 & 0 & 0 & 0 & f & a & c & e \\ 0 & 0 & 0 & 0 & -a & c & -e & -f \\ 0 & 0 & 0 & 0 & c & e & -f & -a \end{bmatrix} \begin{bmatrix} x(0)+x(1)+x(2)+x(3)+x(4)+x(5)+x(6)+x(7) \\ x(0)-x(1)-x(2)+x(3)+x(4)-x(5)-x(6)+x(7) \\ x(0)-x(3)-x(4)+x(7) \\ x(1)-x(2)-x(5)+x(6) \\ x(0)-x(7) \\ x(6)-x(1) \\ x(3)-x(4) \\ x(2)-x(5) \end{bmatrix}$$

Notice then that the $X'(1)$ term is negated to give symmetry to the lower right 4×4 partition. The number of multiplications is now reduced to $0+0+2+2+4+4+4+4=20$.

To further reduce the number of multiplications consider the $X'(2)$ and $X'(6)$ pair:

$$\begin{bmatrix} X'(2) \\ X'(6) \end{bmatrix} = \begin{bmatrix} b & d \\ d & -b \end{bmatrix} \begin{bmatrix} x(0)-x(3)-x(4)+x(7) \\ x(1)-x(2)-x(5)+x(6) \end{bmatrix} \begin{matrix} \leftarrow 2 \text{ multiplications} \\ \leftarrow 2 \text{ multiplications} \end{matrix}$$

requiring four multiplications.

In general, there is a way to eliminate one more multiplication in cases where common matrix elements exist. Assuming

$$\begin{bmatrix} M \\ N \end{bmatrix} = \begin{bmatrix} x & y \\ y & z \end{bmatrix} \begin{bmatrix} m \\ n \end{bmatrix}$$

where M, N, m, n, x and y are simple numbers, we could expand as

$$M = xm + yn \qquad [A.1]$$

$$N = zn + ym \qquad [A.2]$$

Equation [A.1] contains yn but not ym, while [A.2] is the other way round, so we add to each equation its deficient terms:

$$M = xm + yn + \mathbf{ym} - \mathbf{ym} \qquad [A.3]$$

$$N = zn + ym + \mathbf{yn} - \mathbf{yn} \qquad [A.4]$$

By regrouping, $d(m+n)$ is a common product term in both Equations [A.3] and [A.4]:

$$M = y(m+n) + (x-y)m \qquad [A.5]$$

$$N = y(m+n) + (z-y)n \qquad [A.6]$$

Hence only three multiplications are required to compute M and N, and it can be written in the form:

$$\begin{bmatrix} M \\ N \end{bmatrix} = \begin{bmatrix} 1 & 1 & 0 \\ 0 & 1 & 1 \end{bmatrix} \begin{bmatrix} (x-y)m \\ y(m+n) \\ (z-y)n \end{bmatrix}$$

Hence $X'(2)$ and $X'(6)$ can be evaluated as

$$\begin{bmatrix} X'(2) \\ X'(6) \end{bmatrix} = \begin{bmatrix} 1 & 1 & 0 \\ 0 & 1 & 1 \end{bmatrix} \begin{bmatrix} (b-d)\big(x(0)-x(3)-x(4)+x(7)\big) \\ d\big(x(0)+x(1)-x(2)-x(3)-x(4)-x(5)+x(6)+x(7)\big) \\ (-b-d)\big(x(1)-x(2)-x(5)+x(6)\big) \end{bmatrix}$$

or

$$\begin{bmatrix} X'(2) \\ X'(6) \end{bmatrix} = \begin{bmatrix} 1 & 1 & 0 \\ 0 & 1 & -1 \end{bmatrix} \begin{bmatrix} (b-d)\big(x(0)-x(3)-x(4)+x(7)\big) \\ d\big(x(0)+x(1)-x(2)-x(3)-x(4)-x(5)+x(6)+x(7)\big) \\ (b+d)\big(x(1)-x(2)-x(5)+x(6)\big) \end{bmatrix} \begin{matrix} \leftarrow \text{1 multiplication} \\ \leftarrow \text{1 multiplication} \\ \leftarrow \text{1 multiplication} \end{matrix}$$

The remaining four X' terms are:

$$\begin{bmatrix} X'(7) \\ X'(5) \\ -X'(1) \\ X'(3) \end{bmatrix} = \begin{bmatrix} e & f & -a & c \\ f & a & c & e \\ -a & c & -e & -f \\ c & e & -f & -a \end{bmatrix} \begin{bmatrix} x(0)-x(7) \\ x(6)-x(1) \\ x(3)-x(4) \\ x(2)-x(5) \end{bmatrix}$$

Note that the 4×4 matrix can be partitioned as follows:

$$\begin{bmatrix} \begin{bmatrix} X'(7) \\ X'(5) \end{bmatrix} \\ \begin{bmatrix} -X'(1) \\ X'(3) \end{bmatrix} \end{bmatrix} = \begin{bmatrix} \begin{bmatrix} e & f \\ f & a \end{bmatrix} & \begin{bmatrix} -a & c \\ c & e \end{bmatrix} \\ \begin{bmatrix} -a & c \\ c & e \end{bmatrix} & -\begin{bmatrix} e & f \\ f & a \end{bmatrix} \end{bmatrix} \begin{bmatrix} \begin{bmatrix} x(0)-x(7) \\ x(6)-x(1) \end{bmatrix} \\ \begin{bmatrix} x(3)-x(4) \\ x(2)-x(5) \end{bmatrix} \end{bmatrix} \rightarrow \text{4 matrix multiplications}$$

This is of similar form as that of $X'(2)$ and $X'(6)$. Hence we write

$$\begin{bmatrix} \begin{bmatrix} X'(7) \\ X'(5) \end{bmatrix} \\ \begin{bmatrix} -X'(1) \\ X'(3) \end{bmatrix} \end{bmatrix} = \begin{bmatrix} I & I & 0 \\ 0 & I & -I \end{bmatrix} \begin{bmatrix} \begin{bmatrix} e+a & f-c \\ f-c & a-e \end{bmatrix} \begin{bmatrix} x(0)-x(7) \\ x(6)-x(1) \end{bmatrix} \\ \begin{bmatrix} -a & c \\ c & e \end{bmatrix} \begin{bmatrix} x(0)-x(7)+x(3)-x(4) \\ x(6)-x(1)+x(2)-x(5) \end{bmatrix} \\ \begin{bmatrix} e-a & f+c \\ f+c & a+e \end{bmatrix} \begin{bmatrix} x(3)-x(4) \\ x(2)-x(5) \end{bmatrix} \end{bmatrix} \rightarrow \text{3 matrix multiplications}$$

where I is the identity matrix and 0 is the null matrix:

$$I = \begin{bmatrix} 1 & 0 \\ 0 & 1 \end{bmatrix}$$

and

$$0 = \begin{bmatrix} 0 & 0 \\ 0 & 0 \end{bmatrix}$$

Note that each matrix element in the last column is of similar form, hence we can arrive at:

$$\begin{bmatrix} \begin{bmatrix} X'(7) \\ X'(5) \end{bmatrix} \\ \begin{bmatrix} -X'(1) \\ X'(3) \end{bmatrix} \end{bmatrix} = \begin{bmatrix} I & I & 0 \\ 0 & I & -I \end{bmatrix} \begin{bmatrix} \begin{bmatrix} 1 & 1 & 0 \\ 0 & 1 & 1 \end{bmatrix} \begin{bmatrix} (e+a-f+c)(x(0)-x(7)) \\ (f-c)(x(0)-x(7)+x(6)-x(1)) \\ (a-e-f+c)(x(6)-x(1)) \end{bmatrix} \\ \begin{bmatrix} 1 & 1 & 0 \\ 0 & 1 & 1 \end{bmatrix} \begin{bmatrix} (-a-c)(x(0)-x(7)+x(3)-x(4)) \\ c(x(0)-x(7)+x(3)-x(4)+x(6)-x(1)+x(2)-x(5)) \\ (e-c)(x(6)-x(1)+x(2)-x(5)) \end{bmatrix} \\ \begin{bmatrix} 1 & 1 & 0 \\ 0 & 1 & 1 \end{bmatrix} \begin{bmatrix} (e-a-f-c)(x(3)-x(4)) \\ (f+c)(x(3)-x(4)+x(2)-x(5)) \\ (a+e-f-c)(x(2)-x(5)) \end{bmatrix} \end{bmatrix} \begin{matrix} \rightarrow \text{3 multiplications} \\ \rightarrow \text{3 multiplications} \\ \rightarrow \text{3 multiplications} \end{matrix}$$

So in total, $X'(7)$, $X'(5)$, $X'(1)$ and $X'(3)$ require $3 \times 3 = 9$ multiplications.

Along with the three multiplications required, the revised DCT requires 12 multiplications.

To reduce the number of additions and subtractions from 24, we note that the column vectors on the right contains pairs which are used repeatedly, creating a series of sums and differences:

$$\begin{aligned} Add(0) &= x(0) + x(7) \\ Add(1) &= x(3) + x(4) \\ Add(2) &= x(1) + x(6) \\ Add(3) &= x(2) + x(5) \\ Add(4) &= x(0) - x(7) \\ Add(5) &= x(6) - x(1) \\ Add(6) &= x(3) - x(4) \\ Add(7) &= x(2) - x(5) \end{aligned}$$

The matrix equation becomes

$$X'(0) = Add(0) + Add(1) + Add(2) + Add(3) \quad \rightarrow \text{no multiplications}$$

$$X'(4) = Add(0) + Add(1) - Add(2) - Add(3) \quad \rightarrow \text{no multiplications}$$

$$\begin{bmatrix} X'(2) \\ X'(6) \end{bmatrix} = \begin{bmatrix} 1 & 1 & 0 \\ 0 & 1 & -1 \end{bmatrix} \begin{bmatrix} (b-d)(Add(0)-Add(1)) \\ d(Add(0)-Add(1)+Add(2)-Add(3)) \\ (b+d)(Add(2)-Add(3)) \end{bmatrix} \rightarrow \text{3 multiplications}$$

$$\begin{bmatrix} \begin{bmatrix} X'(7) \\ X'(5) \end{bmatrix} \\ \begin{bmatrix} -X'(1) \\ X'(3) \end{bmatrix} \end{bmatrix} = \begin{bmatrix} I & I & 0 \\ 0 & I & -I \end{bmatrix} \begin{bmatrix} \begin{bmatrix} 1 & 1 & 0 \\ 0 & 1 & 1 \end{bmatrix} \begin{bmatrix} (e+a-f+c)(Add(4)) \\ (f-c)(Add(4)+Add(5)) \\ (a-e-f+c)(Add(5)) \end{bmatrix} \\ \begin{bmatrix} 1 & 1 & 0 \\ 0 & 1 & 1 \end{bmatrix} \begin{bmatrix} (-a-c)(Add(4)+Add(6)) \\ c(Add(4)+Add(6)+Add(5)+Add(7)) \\ (e-c)(Add(5)+Add(7)) \end{bmatrix} \\ \begin{bmatrix} 1 & 1 & 0 \\ 0 & 1 & 1 \end{bmatrix} \begin{bmatrix} (e-a-f-c)(Add(6)) \\ (f+c)(Add(6)+Add(7)) \\ (a+e-f-c)(Add(7)) \end{bmatrix} \end{bmatrix} \begin{matrix} \rightarrow 3 \text{ multiplications} \\ \rightarrow 3 \text{ multiplications} \\ \rightarrow 3 \text{ multiplications} \end{matrix}$$

Grouping further:

$$\begin{aligned} Add(8) &= Add(0) + Add(1) \\ Add(9) &= Add(0) - Add(1) \\ Add(10) &= Add(2) + Add(3) \\ Add(11) &= Add(2) - Add(3) \\ Add(12) &= Add(4) + Add(6) \\ Add(13) &= Add(5) + Add(7) \\ Add(14) &= Add(9) + Add(11) \\ Add(15) &= Add(4) + Add(5) \\ Add(16) &= Add(12) + Add(13) \\ Add(17) &= Add(6) + Add(7) \\ Add(18) &= Add(8) + Add10) \\ Add(19) &= Add(8) - Add(10) \end{aligned}$$

Hence

$$\begin{aligned} X'(0) &= Add(18) \\ X'(4) &= Add(19) \end{aligned}$$

$$\begin{bmatrix} X'(2) \\ X'(6) \end{bmatrix} = \begin{bmatrix} 1 & 1 & 0 \\ 0 & 1 & -1 \end{bmatrix} \begin{bmatrix} (b-d)Add(9) \\ (d)Add(14) \\ (b+d)Add(11) \end{bmatrix}$$

$$\begin{bmatrix} \begin{bmatrix} X'(7) \\ X'(5) \end{bmatrix} \\ \begin{bmatrix} -X'(1) \\ X'(3) \end{bmatrix} \end{bmatrix} = \begin{bmatrix} I & I & 0 \\ 0 & I & -I \end{bmatrix} \begin{bmatrix} \begin{bmatrix} 1 & 1 & 0 \\ 0 & 1 & 1 \end{bmatrix} \begin{bmatrix} (e+a-f+c)Add(4) \\ (f-c)Add(15) \\ (a-e-f+c)Add(5) \end{bmatrix} \\ \begin{bmatrix} 1 & 1 & 0 \\ 0 & 1 & 1 \end{bmatrix} \begin{bmatrix} (-a-c)Add(12) \\ (c)Add(16) \\ (e-c)Add(13) \end{bmatrix} \\ \begin{bmatrix} 1 & 1 & 0 \\ 0 & 1 & 1 \end{bmatrix} \begin{bmatrix} (e-a-f-c)Add(6) \\ (f+c)Add(17) \\ (a+e-f-c)Add(7) \end{bmatrix} \end{bmatrix}$$

The previous equations contain 12 products, whose coefficients are

$$
\begin{aligned}
Coe(0) &= b - d \\
Coe(1) &= d \\
Coe(2) &= b + d \\
Coe(3) &= e + a - f + c \\
Coe(4) &= f - c \\
Coe(5) &= a - e - f + c \\
Coe(6) &= -a - c \\
Coe(7) &= c \\
Coe(8) &= e - c \\
Coe(9) &= e - a - f - c \\
Coe(10) &= f + c \\
Coe(11) &= a + e - f - c
\end{aligned}
$$

The products are given by

$$
\begin{aligned}
M(0) &= Coe(0) * Add(9) \\
M(1) &= Coe(1) * Add(14) \\
M(2) &= Coe(2) * Add(11) \\
M(3) &= Coe(3) * Add(4) \\
M(4) &= Coe(4) * Add(15) \\
M(5) &= Coe(5) * Add(5) \\
M(6) &= Coe(6) * Add(12) \\
M(7) &= Coe(7) * Add(16) \\
M(8) &= Coe(8) * Add(13) \\
M(9) &= Coe(9) * Add(6) \\
M(10) &= Coe(10) * Add(17) \\
M(11) &= Coe(11) * Add(7)
\end{aligned}
$$

$$
\begin{aligned}
X'(0) &= Add(18) \\
X'(4) &= Add(19) \\
X'(2) &= M(0) + M(1) \\
X'(6) &= M(1) - M(2)
\end{aligned}
$$

$$
\begin{bmatrix} \begin{bmatrix} X'(7) \\ X'(5) \end{bmatrix} \\ \begin{bmatrix} -X'(1) \\ X'(3) \end{bmatrix} \end{bmatrix} = \begin{bmatrix} I & I & 0 \\ 0 & I & -I \end{bmatrix} \begin{bmatrix} \begin{bmatrix} M(3) + M(4) \\ M(4) + M(5) \end{bmatrix} \\ \begin{bmatrix} M(6) + M(7) \\ M(7) + M(8) \end{bmatrix} \\ \begin{bmatrix} M(9) + M(10) \\ M(10) + M(11) \end{bmatrix} \end{bmatrix}
$$

Note that $M(6) + M(7)$ and $M(7) + M(8)$ will be used twice, hence we define these terms as

$$PostAdd1 = M(6) + M(7)$$
$$PostAdd2 = M(7) + M(8)$$

The final result becomes

$$X'(0) = Add(18)$$
$$X'(4) = Add(19)$$
$$X'(2) = M(0) + M(1)$$
$$X'(6) = M(1) - M(2)$$
$$X'(7) = M(3) + M(4) + PostAdd1$$
$$X'(5) = M(4) + M(5) + PostAdd2$$
$$X'(1) = M(9) + M(10) - PostAdd1 \text{ (note the sign reversal)}$$
$$X'(3) = PostAdd2 - M(10) - M(11)$$

which in matrix form is given as

$$\begin{bmatrix} X'(0) \\ X'(1) \\ X'(2) \\ X'(3) \\ X'(4) \\ X'(5) \\ X'(6) \\ X'(7) \end{bmatrix} = \begin{bmatrix} Add(18) \\ M(9) + M(10) - PostAdd1 \\ M(0) + M(1) \\ PostAdd2 - M(10) - M(11) \\ Add(19) \\ M(4) + M(5) + PostAdd2 \\ M(1) - M(2) \\ M(3) + M(4) + PostAdd1 \end{bmatrix}$$

Along with the 20 addition/subtraction operations in deriving $Add(0) \ldots Add(19)$, the total number of addition/subtraction operations is 32.

The coefficients are calculated based on the given values of a, b, c, d, e and f. They are provided in Table A.4 for easy reference.

The third row gives an integral value scaled by 2^{12} and is used to perform integer multiplication. The actual product can be easily recovered by shifting the respective products 12 bits to the right.

Coeff	Coe(0)	Coe(1)	Coe(2)	Coe(3)	Coe(4)	Coe(5)	Coe(6)	Coe(7)	Coe(8)	Coe(9)	Coe(10)	Coe(11)
Value	0.7654	0.5412	1.8478	2.0531	−0.3902	1.5013	−2.5629	1.1759	−0.9	−3.0727	1.96157	−0.29863
Value *2^12	3135	2217	7568	8410	−1598	6149	−10498	4816	−3686	−12586	8035	−1223

Table A.4 Coefficients 0–11

A.2 1-D Inverse DCT (IDCT)

Consider again the 8×1 column vector with elements $x(0)$, $x(1)$, $x(2)$, $x(3)$, $x(4)$, $x(5)$, $x(6)$ and $x(7)$ whose discrete cosine transform coefficients are $X(0)$, $X(1)$, $X(2)$, $X(3)$, $X(4)$, $X(5)$, $X(6)$ and $X(7)$. The inverse discrete cosine transform equation is

$$X(i) = \frac{1}{2}\sum_{k=0}^{7} C(k)X(k)\cos\left[\frac{(2i+1)k\pi}{16}\right], \quad i = 0, 1, \ldots, 7$$

where

$$C(k) = \begin{cases} \dfrac{1}{\sqrt{2}} & \text{if } k = 0 \\ 1 & \text{otherwise} \end{cases}$$

Note that unlike the DCT, the $C(k)$ terms in the IDCT are inside the summation.

Once again, we replace $C(k)$ by $C'(k)$:

$$X(i) = \frac{1}{2\sqrt{2}}\sum_{k=0}^{7} C'(k)X(k)\cos\left[\frac{(2i+1)k\pi}{16}\right], \quad i = 0, 1, \ldots, 7$$

where

$$C'(k) = \begin{cases} 1 & \text{if } k = 0 \\ \sqrt{2} & \text{otherwise} \end{cases}$$

Again, ignoring the common factors $1/2\sqrt{2}$ we have $x'(i)$ where

$$X(i) = \frac{1}{2\sqrt{2}}x'(i)$$

and

$$x'(k) = \sum_{k=0}^{7} X(k)C'(k)\cos\left[\frac{(2i+1)k\pi}{16}\right], \quad i = 0, 1, \ldots, 7$$

Taking a closer look at the DCT and IDCT equations, it is not hard to derive the matrix multiplication of the IDCT from the DCT as

$$\begin{bmatrix} x'(0) \\ x'(1) \\ x'(2) \\ x'(3) \\ x'(4) \\ x'(5) \\ x'(6) \\ x'(7) \end{bmatrix} = \begin{bmatrix} 1 & 1 & b & d & e & f & -a & c \\ 1 & -1 & d & -b & -f & -a & -c & -e \\ 1 & -1 & -d & b & c & e & -f & -a \\ 1 & 1 & -b & -d & -a & c & -e & -f \\ 1 & 1 & -b & -d & a & -c & e & f \\ 1 & -1 & -d & b & -c & -e & f & a \\ 1 & -1 & d & -b & f & a & c & e \\ 1 & 1 & b & d & -e & -f & a & -c \end{bmatrix} \begin{bmatrix} X(0) \\ X(4) \\ X(2) \\ X(6) \\ X(7) \\ X(5) \\ -X(1) \\ X(3) \end{bmatrix}$$

The IDCT transform matrix is just the transpose of the DCT transform matrix.

By shifting or swapping some rows of the input and its corresponding rows in the transform matrix, we achieve the desired pattern:

$$\begin{bmatrix} x'(0) \\ x'(6) \\ x'(3) \\ x'(2) \\ x'(7) \\ x'(1) \\ x'(4) \\ x'(5) \end{bmatrix} = \begin{bmatrix} 1 & 1 & b & d & e & f & -a & c \\ 1 & -1 & d & -b & f & a & c & e \\ 1 & 1 & -b & -d & -a & c & -e & -f \\ 1 & -1 & -d & b & c & e & -f & -a \\ 1 & 1 & b & d & -e & -f & a & -c \\ 1 & -1 & d & -b & -f & -a & -c & -e \\ 1 & 1 & -b & -d & a & -c & e & f \\ 1 & -1 & -d & b & -c & -e & f & a \end{bmatrix} \begin{bmatrix} X(0) \\ X(4) \\ X(2) \\ X(6) \\ X(7) \\ X(5) \\ -X(1) \\ X(3) \end{bmatrix}$$

This can be decomposed as follows:

$$\begin{bmatrix} x'(0) \\ x'(6) \\ x'(3) \\ x'(2) \\ x'(7) \\ x'(1) \\ x'(4) \\ x'(5) \end{bmatrix} = \begin{bmatrix} \begin{bmatrix} \begin{bmatrix} 1 & 1 \\ 1 & -1 \end{bmatrix}\begin{bmatrix} X(0) \\ X(4) \end{bmatrix} + \begin{bmatrix} b & d \\ d & -b \end{bmatrix}\begin{bmatrix} X(2) \\ X(6) \end{bmatrix} \\ \begin{bmatrix} 1 & 1 \\ 1 & -1 \end{bmatrix}\begin{bmatrix} X(0) \\ X(4) \end{bmatrix} - \begin{bmatrix} b & d \\ d & -b \end{bmatrix}\begin{bmatrix} X(2) \\ X(6) \end{bmatrix} \end{bmatrix} + \begin{bmatrix} e & f & -a & c \\ f & a & c & e \\ -a & c & -e & -f \\ c & e & -f & -a \end{bmatrix}\begin{bmatrix} X(7) \\ X(5) \\ -X(1) \\ X(3) \end{bmatrix} \\ \begin{bmatrix} \begin{bmatrix} 1 & 1 \\ 1 & -1 \end{bmatrix}\begin{bmatrix} X(0) \\ X(4) \end{bmatrix} + \begin{bmatrix} b & d \\ d & -b \end{bmatrix}\begin{bmatrix} X(2) \\ X(6) \end{bmatrix} \\ \begin{bmatrix} 1 & 1 \\ 1 & -1 \end{bmatrix}\begin{bmatrix} X(0) \\ X(4) \end{bmatrix} - \begin{bmatrix} b & d \\ d & -b \end{bmatrix}\begin{bmatrix} X(2) \\ X(6) \end{bmatrix} \end{bmatrix} - \begin{bmatrix} e & f & -a & c \\ f & a & c & e \\ -a & c & -e & -f \\ c & e & -f & -a \end{bmatrix}\begin{bmatrix} X(7) \\ X(5) \\ -X(1) \\ X(3) \end{bmatrix} \end{bmatrix}$$

The matrix is made up of three building blocks:

$$\begin{bmatrix} 1 & 1 \\ 1 & -1 \end{bmatrix}\begin{bmatrix} X(0) \\ X(4) \end{bmatrix} = \begin{bmatrix} X(0)+X(4) \\ X(0)-X(4) \end{bmatrix}$$

$$\begin{bmatrix} b & d \\ d & -b \end{bmatrix}\begin{bmatrix} X(2) \\ X(6) \end{bmatrix} = \begin{bmatrix} 1 & 1 & 0 \\ 0 & 1 & -1 \end{bmatrix}\begin{bmatrix} (b-d)X(2) \\ d[X(2)+X(6)] \\ (b+d)X(6) \end{bmatrix}$$

$$\begin{bmatrix} e & f & -a & c \\ f & a & c & e \\ -a & c & -e & -f \\ c & e & -f & -a \end{bmatrix}\begin{bmatrix} X(7) \\ X(5) \\ -X(1) \\ X(3) \end{bmatrix} = \begin{bmatrix} I & I & 0 \\ 0 & I & I \end{bmatrix}\begin{bmatrix} \begin{bmatrix} 1 & 1 & 0 \\ 0 & 1 & 1 \end{bmatrix}\begin{bmatrix} (e+a-f+c)X(7) \\ (f-c)[X(7)+X(5)] \\ (a-e-f+c)X(5) \end{bmatrix} \\ \begin{bmatrix} 1 & 1 & 0 \\ 0 & 1 & 1 \end{bmatrix}\begin{bmatrix} (-a-c)[X(7)-X(1)] \\ c[X(7)-X(1)+X(5)+X(3)] \\ (e-c)[X(5)+X(3)] \end{bmatrix} \\ \begin{bmatrix} 1 & 1 & 0 \\ 0 & 1 & 1 \end{bmatrix}\begin{bmatrix} (e-a-f-c)[-X(1)] \\ (f+c)[-X(1)+X(3)] \\ (a+e-f-c)X(3) \end{bmatrix} \end{bmatrix}$$

We can reuse all 12 coefficients used in DCT:

$$Coe(0) = b - d$$
$$Coe(1) = d$$
$$Coe(2) = b + d$$
$$Coe(3) = e + a - f + c$$
$$Coe(4) = f - c$$

$$
\begin{aligned}
Coe(5) &= a - e - f + c \\
Coe(6) &= -a - c \\
Coe(7) &= c \\
Coe(8) &= e - c \\
Coe(9) &= e - a - f - c \\
Coe(10) &= f + c \\
Coe(11) &= a + e - f - c
\end{aligned}
$$

The initial sums and differences required are

$$
\begin{aligned}
Add(0) &= X(7) + X(5) \\
Add(1) &= X(7) - X(1) \\
Add(2) &= X(3) + X(5) \\
Add(3) &= X(0) + X(4) \\
Add(4) &= X(2) + X(6) \\
Add(5) &= X(3) - X(1) \\
Add(6) &= Add(0) + Add(5) \\
Add(7) &= X(0) - X(4)
\end{aligned}
$$

The 12 multiplications are

$$
\begin{aligned}
M(0) &= Coe(0) * X(2) \\
M(1) &= Coe(1) * Add(4) \\
M(2) &= Coe(2) * X(6) \\
M(3) &= Coe(3) * X(7) \\
M(4) &= Coe(4) * Add(0) \\
M(5) &= Coe(5) * X(5) \\
M(6) &= Coe(6) * Add(1) \\
M(7) &= Coe(7) * Add(6) \\
M(8) &= Coe(8) * Add(2) \\
M(9) &= Coe(9) * X(1) \\
M(10) &= Coe(10) * Add(5) \\
M(11) &= Coe(11) * X(3)
\end{aligned}
$$

The building blocks expressed in the sums and products are

$$
\begin{bmatrix} 1 & 1 \\ 1 & -1 \end{bmatrix} \begin{bmatrix} X(0) \\ X(4) \end{bmatrix} = \begin{bmatrix} Add(3) \\ Add(7) \end{bmatrix}
$$

$$
\begin{bmatrix} b & d \\ d & -b \end{bmatrix} \begin{bmatrix} X(2) \\ X(6) \end{bmatrix} = \begin{bmatrix} M(0) + M(1) \\ M(1) - M(2) \end{bmatrix}
$$

$$\begin{bmatrix} e & f & -a & c \\ f & a & c & e \\ -a & c & -e & -f \\ c & e & -f & -a \end{bmatrix} \begin{bmatrix} X(7) \\ X(5) \\ -X(1) \\ X(3) \end{bmatrix} = \begin{bmatrix} I & I & 0 \\ 0 & I & I \end{bmatrix} \begin{bmatrix} \begin{bmatrix} M(3)+M(4) \\ M(4)+M(5) \end{bmatrix} \\ \begin{bmatrix} M(6)+M(7) \\ M(7)+M(8) \end{bmatrix} \\ \begin{bmatrix} -M(9)+M(10) \\ M(10)+M(11) \end{bmatrix} \end{bmatrix} = \begin{bmatrix} M(3)+M(4)+M(6)+M(7) \\ M(4)+M(5)+M(7)+M(8) \\ M(6)+M(7)-M(9)+M(10) \\ M(6)+M(7)+M(10)+M(11) \end{bmatrix}$$

Hence the IDCT can be simplified as

$$\begin{bmatrix} x'(0) \\ x'(6) \\ x'(3) \\ x'(2) \\ x'(7) \\ x'(1) \\ x'(4) \\ x'(5) \end{bmatrix} = \begin{bmatrix} [Add(3)]+[M(0)+M(1)]+[M(3)+M(4)+M(6)+M(7)] \\ [Add(7)]+[M(1)-M(2)]+[M(4)+M(5)+M(7)+M(8)] \\ [Add(3)]-[M(0)+M(1)]+[M(6)+M(7)-M(9)+M(10)] \\ [Add(7)]-[M(1)-M(2)]+[M(6)+M(7)+M(10)+M(11)] \\ [Add(3)]+[M(0)+M(1)]-[M(3)+M(4)+M(6)+M(7)] \\ [Add(7)]+[M(1)-M(2)]-[M(4)+M(5)+M(7)+M(8)] \\ [Add(3)]-[M(0)+M(1)]-[M(6)+M(7)-M(9)+M(10)] \\ [Add(7)]-[M(1)-M(2)]-[M(6)+M(7)+M(10)+M(11)] \end{bmatrix}$$

Since both the DCT and IDCT use the same coefficients and matrix blocks, it is also economical in a hardware implementation as well.

A.3 Application in 2-D DCT

The 2-D DCT is given by

$$X(m,n) = \frac{1}{4}C(m)C(n)\sum_{j=0}^{7}\sum_{i=0}^{7}x(i,j)\cos\left[\frac{(2i+1)m\pi}{16}\right]\cos\left[\frac{(2j+1)n\pi}{16}\right], \quad m,n=0,1,\ldots,7$$

where

$$C(k) = \begin{cases} \dfrac{1}{\sqrt{2}} & \text{if } k=0 \\ 1 & \text{otherwise} \end{cases} \quad \text{for } m,n=0,1,\ldots,7$$

As DCT is separable, we could perform the optimised 1-D DCT column-wise followed by row-wise. As for the 1-D case we shall use the modified version of $C'(k)$:

$$X(m,n) = \frac{1}{8}C'(m)C'(n)\sum_{j=0}^{7}\sum_{i=0}^{7}x(i,j)\cos\left[\frac{(2i+1)m\pi}{16}\right]\cos\left[\frac{(2j+1)n\pi}{16}\right], \quad m,n=0,1,\ldots,7$$

where

$$C'(k) = \begin{cases} 1 & \text{if } k=0 \\ \sqrt{2} & \text{otherwise} \end{cases} \quad \text{for } m,n,=0,1,\ldots,7$$

Hence, after a column-wise optimised 1-D DCT and a row-wise 1-D DCT, the resulting values are right shifted by three bits (equivalent to a division by 8) to obtain the $X(m,n)$. (Recall that the factor $\frac{1}{2}r_2$ has been ignored in section A.2.)

Appendix B

Block memory display

To be able to display the content of a memory block from the EVM can be very useful in checking large amounts of data, especially when the data represent images. Two methods are used in this book. The first relies on the PC to display the data, which is transferred from the DSP via the PCI interface (see Chapter 9). The second method is much simpler since it involves saving memory directly to a file. However, it cannot be used for real-time display. The procedure for such a method is described below:

(1) Run the DSP program.
(2) Once the DSP program has been executed, save the data memory block of interest using the 'Save' command available under the window 'Memory' button as shown in Figure B.1, where the `image_in` and `image_out` are the labels used for the addresses of the input and processed data respectively. The 'Length' represents the number of words to be saved (with the DCT and IDCT the length was 256×256 bytes $=$ 16,384 words).

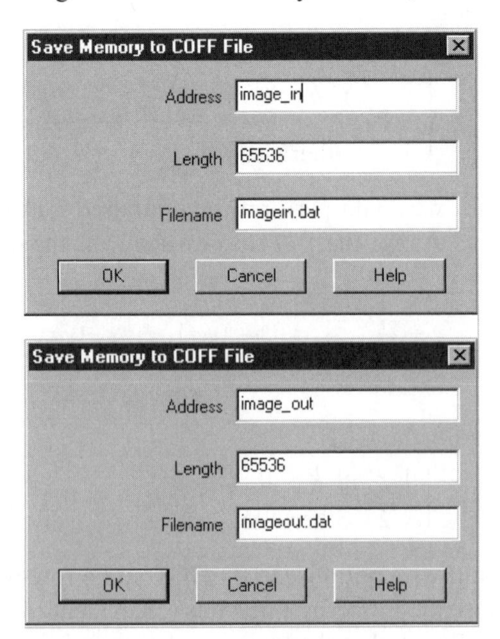

Figure B.1 Screen captures from the EVM debugger

(a)

(b)

Figure B.2 Resulting display of the two blocks: (a) from `image_in.dat`; (b) from `image_out.dat`

(3) Run the `display2.m` file from MATLAB shown in Program B.1. This file takes two files as input data (`imagein.dat` and `imageout.dat`) and displays them on the screen. Program B.1 can be modified to take more input files if required. The output produced by the `display.m` files when using the program `F:\DSPCODE\DCT\SLOWDCT\DCT.OUT` is shown in Figure B.2.

```
cmap = zeros (256,3);
for i=0:255,
  cmap(i+1,1) = i/255;
  cmap(i+1,2) = i/255;
  cmap(i+1,3) = i/255;
end

FID1=fopen('imagein.dat');
FID2=fopen('imageout.dat');
% FID3=fopen('imageouto3.dat');

[dummy, count]=fread (FID1,[1,70],'uchar');
[dummy, count]=fread (FID2,[1,70],'uchar');
%[dummy, count]=fread (FID3,[1,70],'uchar');

[T1, count]=fread(FID1,[256,256],'uchar');
[T2, count]=fread(FID2,[256,256],'uchar');
%[T3, count]=fread(FID3,[256,256],'uchar');

fclose(FID1);
fclose(FID2);
%fclose(FID3);

figure(1)
image(T1')

colormap(cmap);

figure(2)
image(T2')
colormap(cmap);

%figure(3)
%image(T3')
%colormap(cmap);
```

Program B.1 display2.m file for displaying block of memory

References

Ahmed X.X. *et al.* (1974) Discrete Cosine Transform, *IEEE Trans. Comput.*, pp. 90–93, January
Bozic S.M. (1994) *Digital and Kalman Filtering*, Edward Arnold.
Haykin S. (1996) *Adaptive Filter Theory*, Prentice Hall.
Ifeachor E.C. and Jervis B.W. (1993) *Digital Signal Processing – A Practical Approach*, Addison-Wesley.
Ludeman C.L. (1987) *Fundamentals of Digital Signal Processing*, John Wiley.
Mattison P.E. (1994) *Practical Digital Video with Programming Examples in C*, John Wiley.
McClellan J.H., Schafer R.W. and Yoder M.A. (1998) *DSP First – A Multimedia Approach*, Prentice Hall.
McGovern F.A., Woods R.F. and Yan M. (1994) Novel VLSI implementation of (8 × 8) point 2-D DCT, *Electronics Letters*, 14th April, Vol. 30, No. 8.
Mitra S.K. (1998) *Digital Signal Processing – A Computer-Based Approach*, McGraw-Hill.
Parks T.W. and Burrus C.S. (1987) *Digital Filter Design*, John Wiley.
Preaches J.G. and Manolakis D.G. (1992) *Digital Signal Processing – Principles, Algorithms, and Applications*, Macmillan.
SPRU186, *Assembly Language Tools User's Guide*, Texas Instruments.
SPRU187, *Optimizing C Compiler User's Guide*, Texas Instruments.
SPRU188, *C Source Debugger User's Guide*, Texas Instruments.
SPRU189, *CPU and Instruction Set Reference Guide*, Texas Instruments.
SPRU190, *Peripherals Reference Guide*, Texas Instruments.
SPRU198, *Programmer's Guide*, Texas Instruments.
SPRU273, *Peripheral Support Library*, Texas Instruments.
Strum R.D. and Kirk D.E. (1988) *First Principles of Digital Signal Processing*, Addison-Wesley.
Widrow B. and Stearns S.D. (1985) *Adaptive Signal Processing*, Prentice Hall.

Further reading

Baudoin G. and Virolleau F. (1996) *DSP les Processeurs de Traitement du Signal – Famille 320C5X*, Dunos.
Chassaing R. (1990) *Digital Signal Processing with C and the TMS320C25*, John Wiley.
Chassaing R. (1992) *Digital Signal Processing with C and the TMS320C30*, John Wiley.
Chen J. and Sorensen H.V. (1997) *A Digital Signal Processing Laboratory using the TMS320C30*, Prentice Hall.
Davies P. (1995) *The Indispensable Guide to C*, Addison-Wesley.
Deller J.R. and Grover D. (1999) *Digital Signal Processing – and the Microcontroller*, Prentice Hall.
Embree P.M. (1995) *C Algorithm for Real-Time DSP*, Prentice Hall.
Hahn D.B. (1997) *Essential MATLAB – for Scientists and Engineers*, Arnold.
Higgins R.J. (1990) *Digital Signal Processing in VLSI*, Prentice Hall.
Ingle V.K. and Proakis J.G. (1997) *Digital Signal Processing – Using MATLAB V.4*, PWS.

References

Proceedings of the First European DSP Education and Research Conference, Texas Instruments, September 1996.

Proceedings of the Second European DSP Education and Research Conference, Texas Instruments, September 1998.

Tretter S.A. (1995) *Communication System Design using DSP Algorithms – With Laboratory Experiments for the TMS320C30*, Plenum Press.

Index